TRANSHUMANIST DREAMS AND
DYSTOPIAN NIGHTMARES

Transhumanist Dreams AND ⌒
Dystopian Nightmares

The Promise and Peril of Genetic Engineering

MAXWELL J. MEHLMAN

The Johns Hopkins University Press

Baltimore

© 2012 The Johns Hopkins University Press
All rights reserved. Published 2012
Printed in the United States of America on acid-free paper
9 8 7 6 5 4 3 2 1

The Johns Hopkins University Press
2715 North Charles Street
Baltimore, Maryland 21218-4363
www.press.jhu.edu

Library of Congress Cataloging-in-Publication Data

Mehlman, Maxwell J.
    Transhumanist dreams and dystopian nightmares : the promise and peril of genetic
engineering / Maxwell J. Mehlman.
        p. cm.
    Includes bibliographical references and index.
    ISBN 978-1-4214-0669-5 (hbk. : acid-free paper) —
    ISBN 1-4214-0669-1 (hbk. : acid-free paper) —
    ISBN 978-1-4214-0727-2 (electronic) —
    ISBN 1-4214-0727-2 (electronic)
    1. Genetic engineering—Social aspects. 2. Genetic engineering—Moral and ethical
aspects. 3. Genetic engineering—Environmental aspects. I. Title.
    QH442.M44 2012
    660.6'5—dc23   2012001949

A catalog record for this book is available from the British Library.

*Special discounts are available for bulk purchases of this book. For more information,
please contact Special Sales at 410-516-6936 or specialsales@press.jhu.edu.*

The Johns Hopkins University Press uses environmentally friendly book materials,
including recycled text paper that is composed of at least 30 percent post-consumer
waste, whenever possible.

For my children's children

CONTENTS

ACKNOWLEDGMENTS

Support for the research on this book was provided by the Metanexus and Templeton Foundations under a grant from the Center for the Study of Religion and Conflict, Arizona State University; by a grant to the center for Genetic Research Ethics and Law, Case Western Reserve University, by the Ethical, Legal, and Social Implications Program of the National Human Genome Research Institute, National Institutes of Health (1 P50 HG03390-01); and by the Center for Health Law Studies at St. Louis University School of Law. The author also would like to thank Cory Schmidt and Kelsey Marand for their research assistance and Wendy Harris, Suzanne Flinchbaugh, and Jeremy Horsefield for their excellent editing.

TRANSHUMANIST DREAMS AND
DYSTOPIAN NIGHTMARES

Introduction

According to a recent Gallup Poll, 80 percent of Americans reject the theory of natural evolution. Some of them take a more literal view of the Bible and believe that God made humans pretty much in their present form within the past 10,000 years, while others are willing to accept scientific evidence of evolution but think that God actively guides evolutionary processes.[1] No doubt one of the reasons for the popularity of these beliefs is that the notion of natural evolution conjures up a terrifying vision of a universe without meaning or direction in which humanity is accidental and insignificant. "The universe that we observe," evolutionary biologist and atheist Richard Dawkins bluntly insists, "has precisely the properties we should expect if there is, at bottom, no design, no purpose, no evil, no good, nothing but pitiless indifference."[2]

It is true that evolution takes place to some extent by chance. Changes in genes occur more or less at random, these random changes make organisms more or less successful depending on environmental conditions, and the successful ones reproduce, thereby passing the genetic changes on to their descendants. Over time, these cumulative changes alter the way an organism looks and acts. But even within natural evolution, there is an element of choice. The reproductive process is seldom left completely to chance. Females often select their mates, who compete with one another by being bolder, having prettier plumage, swimming faster, building better nests, or, in the case of certain frogs, being more enthusiastic croakers. In some species, males also do some of the selecting. These reproductive choices determine which sets of genes will be inherited by offspring.

Like other animals, humans have developed elaborate behaviors to influence the selection of their mates. Dinka men in the Sudan hold a three-month-long contest to see who can grow fattest by drinking a mixture of milk and cow urine.[3] Wodaabe men in Niger use exaggerated facial expressions to display their charm and good looks in periodic dance competitions; rolling the right eye back and forth is deemed especially attractive.[4] Upper-class American girls may be invited to "come out" at debutante balls, where they are judged on their dancing ability, etiquette, and how well they curtsy.[5]

But humans have gone much further than other animals in avoiding the randomness of reproduction. One recent phenomenon is computer-assisted dating. On the website eHarmony.com, for example, people are matched on the basis of 29 physical and mental characteristics that the company assesses on the basis of answers to an online questionnaire. These characteristics include self-concept, emotional management, obstreperousness, romantic and sexual passion, kindness, dominance, intellect, curiosity, humor, artistic passion, industry, appearance, traditionalism, spirituality, ambition, altruism, family background, family status, and education.[6] A rival online dating company, Match.com, has 1.2 million paid members and annual revenues of over $360 million.[7]

Computer-assisted dating has even begun to make use of genetic testing. For a $2,000 lifetime membership, one company will test a man's DNA for clues to the nature of his immune system and then match him up with women who have very different immune systems, on the basis of research showing that women prefer the smell of men whose immune systems are least like their own.[8] Another company, 23andMe, one of whose creators is married to Google cofounder Sergey Brin, sells testing over the Internet for over a hundred genetic characteristics. These include tests not only for diseases such as age-related macular degeneration, prostate cancer, and bipolar disorder but for nondisease traits such as intelligence and longevity. All you have to do is give the company your credit card number over the Internet, spit into a little bottle that the company mails to you, and send it back in the accompanying self-addressed, stamped envelope. A few weeks later, you go on the Internet and get your results.

This company has an innovative marketing plan. While the usual assumption is that people want to keep their personal genetic information a secret to avoid stigma and discrimination, 23andMe is counting on the younger generation's supposed lack of inhibition to turn personal genetic information into something to be shared with others, such as on Facebook. To facilitate the sharing of test information, the company has launched a community website that gives customers, in the company's words, "the ability to create a profile, connect with others, share stories, ask questions about specific traits and ancestry groups and learn more about research studies."[9] Along the same lines, in 2008, the company sponsored a "spit party" in New York where a group of prominent socialites, including Rupert Murdock, Harvey Weinstein, and Diane von Furstenberg, got together, spit into bottles, and sent them to the company for analysis. It won't be long before companies like 23andMe offer dating services in which they match people up on the basis of their results on multiple genetic tests.

Genetic testing has opened up other possibilities for avoiding random reproduction. Tay-Sachs is a genetic disease prevalent in populations descended from European Jews which afflicts children shortly after they are born and causes their deaths within a few years. If two people who each carry the genetic mutation that causes Tay-Sachs have a child, the child has a 25 percent chance of inheriting the disease. In some Orthodox Jewish communities in which marriages are still arranged, teenagers are tested to see if they have the mutation. The teenagers are not told the result of the test, but the head rabbi is, so that he can refuse to give his consent to a marriage between two carriers. As a result, Tay-Sachs has been all but stamped out in these communities.

Advances in infertility medicine, especially in vitro fertilization (IVF), in which eggs are fertilized in the laboratory rather than in the woman's fallopian tubes (in vitro means "in glass"), have given humans more direct ways of influencing the genes that their children inherit. In preimplantation genetic diagnosis (PGD), cells are removed from the dozen or so embryos that are produced in the lab and subjected to genetic testing, after which the parents can pick the two or three embryos with the best sets of genes to implant in the womb for a chance to come to term. Alternatively, fetuses can be tested in the womb, and those with poor test results can be aborted. In either case, the parents are opting for or against certain sets of genes, and those that are opted against are not passed on to the parents' descendants.

Another infertility treatment that affects the way genes are handed down through generations is artificial insemination. Originally developed to treat infertility, it is also now used to enable people to combine their genes with genes from specially selected donors. There already is a flourishing market for "exceptional" eggs and sperm to be used with artificial insemination. To sell eggs to one company, Fertility Alternatives, you must have graduated from or currently be attending a major university, preferably one in the Ivy League, and have a GPA of over 3.0, SAT scores above 1350, and a high IQ.[10]

Infertility treatments also alter natural evolution by allowing people who cannot reproduce naturally to pass on their genes. Indeed, every time modern medicine keeps a sick person alive long enough to reproduce, it can be said to thwart the process of natural evolution.

Yet a far more radical method for supplanting natural evolution will become available in the not-too-distant future: intentionally modifying inherited genes themselves, known as germ line genetic manipulation. Physicians have taken the first step by introducing functional genes into the bodies of people who lack them. In 1990, they infused modified genes into a 4-year-old girl, Ashanti DeSilva, who had been born without a properly functioning immune system, and the genes began to manufacture the enzyme she was missing. Since then, there have been almost a thousand similar gene therapy efforts. The modified genes in these cases are not passed on to the patients' children, so the main evolutionary effect is that these patients now may live long enough to reproduce. But scientists have learned how to splice genes from one organism into the DNA of another so that future generations also possess the spliced-in genes. They have created bacteria that can biodegrade oil spills, as well as bacteria that mass-produce unlimited quantities of substances found naturally in the human body to treat conditions caused by hormonal deficiencies, such as dwarfism.

It's a long way from engineering the DNA of bacteria to human germ line genetic engineering; genetic engineering in primates lags well behind what can be accomplished in simpler animals. Yet science passed a milestone in 2009 when Japanese researchers took the embryos of marmosets, a species of New World monkey, inserted a jellyfish gene that made the monkeys glow green under ultraviolet light, and found that the monkeys produced babies that glowed just like their parents.[11] Glow-in-the-dark monkeys may not sound like much of an evolutionary step, but this was the first time that the DNA of a primate had been reengineered at a sufficiently early stage of the animal's development so that the changes were inherited by its descendants. Eventually this technique will be perfected in the monkey. No doubt it will then be tried in human embryos. This will open the door to human germ line genetic manipulation.

For the time being, we treat inherited disorders such as cystic fibrosis and hereditary colon cancer with what Lewis Thomas labeled "half-way" technologies.[12] We use surgery, drugs, and dietary changes and are gradually introducing gene therapy like the one Ashanti DeSilva received, but the genes that cause these illnesses are still present in the reproductive cells of sufferers, and therefore their children remain at risk for getting the disease and passing it on to their offspring. Germ line gene therapy would remove the risk by preventing succeeding generations from inheriting the genes that cause the disease. The next step will be to learn which pieces of DNA are associated with nondisease traits and to splice these segments into and out of human DNA in such a way that the resulting changes are passed on to our offspring. At that point, we will have crossed an evolutionary Rubicon. From then on, so long as humans maintain this technological wherewithal, they will be in a position to avoid random genetic change altogether. They will enter a new evolutionary phase. Paleontologist Peter Ward calls it "directed evolution," borrowing a term from protein engineering that refers to a process for developing new proteins quickly by mimicking natural forces.[13] Others use the term "controlled" or "rational" evolution. I call it "evolutionary engineering."

Some people think that evolutionary engineering is inevitable. British biologist Julian Huxley (Aldous's brother) saw it coming in 1957, when he wrote, "It is as if man had been suddenly appointed managing director of the biggest business of all, the business of evolution—appointed without being asked if he wanted it, and without proper warning and preparation. What is more, he can't refuse the job. Whether he wants to or not, whether he is conscious of what he is doing or not, he is in point of fact determining the future direction of evolution on this earth. That is his inescapable destiny, and the sooner he realizes it and starts believing in it, the better for all concerned."[14] Pioneering bioethicist Joseph Fletcher, who was also an ordained Episcopal priest, put it in a historical perspective: "We began our human history by learning to control the physical environment (and still make serious mistakes). We have made some progress in controlling our social life, and we are learning to control our behavior. It is time, then, that we accepted control of our heredity."[15] UCLA biophysicist Gregory Stock declares that "the time has come for us to accept the responsibility that comes with our new understandings and powers."[16] Fletcher goes so far as to view the failure to do so as "immoral."[17]

Some evolutionary biologists regard humans primarily as vessels that genes use to ferry themselves from one generation to the next. They see directed evolution as a chance to turn the tables. Richard Dawkins, the evolutionary biologist who described genes as "selfish," observed that "we are built as gene machines . . . but we have the power to turn against our creators."[18] Oxford philosopher Julian Savulescu asserts that "humanity until this point has been a story of evolution for the survival of genes. Now we are entering a new phase of human evolution—evolution under reason—where human beings are masters of their destiny. Power has been transferred from nature to science."[19]

Some people look forward to the prospect of germ line genetic engineering as an opportunity to reengineer the human species. They call themselves "transhumanists," a term coined by Julian Huxley. According to Nick Bostrum, he and his fellow transhumanists believe that "current human nature is improvable through the use of applied science and other rational methods, which may make it possible to increase human health-span, extend our intellectual and physical capacities, and give us increased control over our own mental states and moods."[20] For transhumanists, directed evolution is the apotheosis of the human species. Stock, for example, announces that "humanity is leaving its childhood and moving into its adolescence as its powers infuse into realms hitherto beyond our reach."[21] Some transhumanists believe that by equipping us to survive in the inhospitable environment of space, evolutionary engineering will be what saves humanity in the event of a planetary catastrophe, like the eventual death of the Sun. Quite literally, it could be our ticket to the stars.

For many religious conservatives, however, evolutionary engineering is a sacrilege. In a speech in 2007 to the members of the Pontifical Academy for Life, Pope Benedict XVI decried "the interest in more refined biotechnological research [which] is growing in the more developed countries in order to establish subtle and extensive eugenic methods, even to obsessive research for the 'perfect child.'"[22] The former chair of President Bush's Council on Bioethics, Leon Kass, is worried that radical changes in human beings would conflict with God's design: "Most of the given bestowals of nature have their given species-specified natures: they are each and all of a given *sort*. . . . We need more than generalized appreciation for nature's gifts. We need a particular regard and respect for the special gift that is our own given nature" (emphasis in original).[23]

The views of Pope Benedict and Leon Kass are those of nonscientists whose belief systems are threatened by directed evolution. What do the scientists who are actually developing these technologies think? Some, pointing to the astounding intricacies of genetics, are afraid that humans will never master genetic engineering. Important aspects will always elude us, they insist, such as manipulating multifactorial traits, which are produced by interactions between and among genes and the environment. Attempts at evolutionary engineering are bound to fail.

Other scientists are not so pessimistic, however. They point to how much we have learned about human genetics since the completion of the Human Genome Project, the multiyear, multi-billion-dollar program to decode human DNA. When the project began in 1990, only around 100 of the 3 billion nucleotides in a sample of human DNA could be processed in a day with the technology that was available. Now, the process is highly automated, and massive amounts of DNA are sequenced simultaneously. Decoding has become not only much faster but more accurate and much less expensive. A recent report for the Department of Defense pointed out that the exponential increase in the ability to sequence DNA has far exceeded "Moore's Law," the 1970 forecast by Intel cofounder Gordon Moore that the number of transistors in computer chips would double every two years, which has held true for over 40 years. It now costs only about $20,000 to decode a person's genome, and this is expected to drop to less than $1,000 by 2012 and to $100 by 2013.[24]

Along with accelerating the pace of genetic sequencing, researchers are becoming more adept at linking genetic information with human characteristics. Francis Collins, the director of the National Institutes of Health (NIH), hails the tremendous progress that has been made in identifying genes containing the instructions for complex human traits, that is, traits that are produced by more than one gene or by a combination of genes and the environment. In 2002, Collins notes, researchers had discovered the genes for only seven of these traits, but six years later, the number had grown to 130.[25] Another scientist who thinks that genetic engineering has made great strides is Theodore Friedman, the former president of the American Society of Gene Therapy. In 1997, he declared that "there are no insurmountable scientific barriers to genetic enhancement,"[26] that is, giving people desirable genetic traits. In view of what experts like Collins and Friedman believe, radical forms of germ line genetic engineering could become a reality much sooner than many people think.

Whether scientists are dubious or optimistic about the prospects for rational evolution, though, they tend to agree on two things. First, however long it will take to perfect the necessary technology, it is inevitable that humans will attempt to control their evolutionary future, and second, in the process of learning how to direct human evolution, we are bound to make mistakes.

This book is about these mistakes. What kinds of mistakes are likely to be made, and how serious will they be? Who will be harmed? Will mistakes be confined to a few people serving as experimental subjects, or will they be widespread? Will their effects be felt immediately or in the longer term? Above all, can we avoid extinguishing humankind itself?

Religious conservatives are not the only critics of engineered evolution; the perils of genetic manipulation inspire numerous fictional accounts of the human future. Chapter 1 contrasts these bleak dystopias with the rapturous enthusiasm of the transhumanists. Genetic engineering is not the first technology that has raised fears of total human extinction, and chapter 2 reviews how similar scares have been handled in the past. The book then delves more specifically into the hazards of evolutionary engineering, considering first how individuals who were engineered might be harmed (chaps. 3 and 4), next how genetic engineering could disrupt cultural progress by undermining the political and social framework of society (chap. 5), and then how it might wipe out humans altogether (chap. 6). Chapter 7 places genetic engineering in the context of what we know about the operation of natural evolution. The discussion concludes by considering how each of these risks might be reduced (chaps. 8–11).

PART ONE

What Is to Come?

CHAPTER ONE

Visions of Heaven and Hell

There seems to be no middle ground when it comes to envisioning a future in which evolutionary engineering is commonplace; it is portrayed as either loathsome or sublime. The transhumanists are the cheerleaders, and some of them get downright giddy when they gaze into their crystal balls. Among the most hyperbolic is Simon Young, who started out as a piano player and composer; taught music at Trinity College of Music in London, where he is now an Emeritus Fellow; and eventually became a licensed massage therapist after he developed back pain while performing.[1] Young describes how he came across the concept of transhumanism in 1984 while reading a book in his father's library and then spent the next 20 years developing "a modern philosophy suitable for the 21st century (I was thinking ahead!)."[2] Eventually he distilled "thousands of pages" of his thoughts into a 417-page book entitled *Designer Evolution: A Transhumanist Manifesto*, which he published in 2006. As Young tells the story, "I developed my philosophical system gradually and methodically over some 20 years, while supporting myself as a piano player in the hotels and piano bars of London and Europe (the term 'Designer Evolution,' for instance, occurred to me while playing Jerome Kern / Fred Astaire's 'Pick Yourself Up,' at The Ritz)."[3] "Transhumanism," raves Young, "is not just the 'Big Idea' of the 21st century, but the biggest idea *in the whole of human history—ever!*" As a result of "the combined miracles of Superbiology," an amalgam of biotechnology and genetics that Young also refers to as "Supergenics," humans will complete their evolution from the Stone Age, through the Iron and Industrial Ages, to the "DNAge, defined by the ability to manipulate human life itself."[4] Humans must seize this opportunity, Young says, because they have "the will to evolve," and they must not be fainthearted. "Let us cast away cowardice," he exhorts, "and seize the torch of Prometheus with both hands."[5]

Although the end point that Young and his fellow transhumanists look forward to is transcendent, they concede that progress will be gradual; indeed, it must be, so that people can become accustomed to the powerful new biological tools at their command. The transformation of humanity has already begun with egg and sperm donation, which allows parents to select donors who have favorable sets of characteristics that the parents hope will be incorporated into their offspring. IVF and genetic testing enable parents to select and implant only those embryos that possess the best genes. More active forms of genetic engineering lie just around the corner, transhumanists point out, with the somatic type of gene therapy that gave Ashanti DeSilva a partially functioning immune system to be followed by genetic manipulation that eradicates the disorder from the germ line. People may be alarmed at first by these new technological capabilities, transhumanists admit, but eventually they will get used to them. "Germline engineering represents a shift in human reproduction," admits Gregory Stock, "but as effective somatic therapies become common, reduced public concern about genetic interventions in general will smooth the way for a move from screening and selecting embryos to actually manipulating them."[6]

Transhumanists also accept the fact that active forms of evolutionary engineering will begin by targeting disease. According to transhumanist James Hughes, "the first beneficiaries of these technologies will be the sick and disabled, for there is little controversy that they should be able to use technology to more fully control their own lives."[7] Not all diseases will be targeted at once, observes Princeton molecular biologist Lee Silver: "It will begin in a way that is most ethically acceptable to the largest portion of society, with the treatment of only those childhood diseases like sickle cell anemia and cystic fibrosis—that have a severe impact on quality of life. The number of parents who will desire this service will be tiny, but their experience will help to ease society's trepidation."[8]

Gene therapy for serious childhood diseases will be followed by genetic treatments for diseases that are less severe for children or that do not manifest symptoms until adulthood, such as predispositions to obesity, diabetes, heart disease, asthma, and some forms of cancer.[9] One technique for treating cancer that Gregory Stock proposes is inserting a gene into cells which instructs them to manufacture a poison that kills them, but which only switches on in the presence of a hormone that is not naturally present in the body. A doctor would inject the hormone into a cancer patient at the site of a tumor, stimulating the cancer cells to release the suicidal poison.[10] Stock also is excited about artificially constructed chromosomes, which could transport genes that combat cancer, aging, and obesity into the body. Human artificial chromosomes or

HACs were developed in 1997; the idea is to embed desired genes in small synthetic chromosomes that can be introduced into cell nuclei, where they would take up residence along with the other naturally occurring chromosomes and be passed on to daughter cells when the parent cells divided.

Another approach that transhumanists have embraced is a genetic vaccine. John Harris, a law professor and bioethicist at the University of Manchester in England, for example, describes how David Baltimore's lab at Caltech is exploring the possibility of engineering resistance to cancer and viruses such as HIV by manipulating the immune system, in much the same way that a vaccine stimulates the production of antibodies.[11] This project is "a grand challenge," says Harris, and "the benefits of this and related work around the world are incalculable."[12] Former *Washington Post* reporter Joel Garreau asks why this vaccine idea could not be extended to pain, so that people felt acute pain only as long as it took to make it clear to them that something was wrong and that they needed to obtain medical help.[13]

After physical disease and unnecessary pain have been overcome, transhumanists turn their attention to mental ailments. "The final frontier," proclaims Lee Silver, "will be the mind and the senses."[14] Microsoft computer scientist Ramez Naam, author of the 2005 book *More Than Human: Embracing the Promise of Biological Enhancement*, predicts that ADHD, bipolar disorder, depression, and anxiety will soon fall to the assaults of the gene doctors.[15] Silver adds alcoholism to this list, while Stock tosses in drug addiction. "Between gene therapies, better drugs and nano-neuro brain prostheses," James Hughes predicts, "mental illness will likely join cancer and aging in being completely preventable and controllable by the latter twenty-first century."[16] For that matter, asks Simon Young, why not proceed to correct emotional dysfunction that falls short of full-blown mental illness, such as nervousness, gloominess, timidity, and lethargy?[17]

With humans now healthy both physically and mentally, the next item on the transhumanists' agenda is genetic enhancement. Young foresees that, like the physical diseases that gene therapy will tackle first, the first traits that will be enhanced genetically will be physical characteristics such as strength, stamina, vitality, and virility.[18] Physical beauty certainly will be a prime candidate. In 2010, 11 million American women and more than a million American men paid $10 billion for 12.5 million cosmetic procedures, including almost 300,000 breast enlargements, 250,000 nose jobs, 230,000 eyelid procedures, 200,000 liposuctions, 115,000 tummy tucks, and 5 million Botox treatments.[19] More than 200,000 of these consumers were under the age of 19.[20] Asians who prefer the look of Western eyes undergo eyelid surgery to add a crease in the upper lid. In China, where being tall is a premium for certain high-paying jobs, people endure leg-lengthening surgery in which their shin bones are repeatedly sawn through, pulled apart, and fixed in place with metal pins and braces until the ends of the bone attach to each other. If people are willing to go to these lengths to improve their appearance, there is no doubt that parents will seek to use genetic modification to improve their children's looks.

Another popular type of modification will be enhancing physical performance. The constant battles between sports officials and rogue athletes, coaches, trainers, and chemists testify to the huge demand for biomedical advantages in sports. In 2005, the U.S. Anti-Doping Agency conducted 8,175 tests for banned substances and imposed sanctions on 20 athletes.[21] The World Anti-Doping Agency (WADA) performed 3,279 tests worldwide in 2006, 57 of which were positive.[22] WADA even runs testing programs for the Paralympics, the competition for elite athletes with disabilities that is held alongside the regular Olympics. In the 2004 Athens Paralympic Games, one powerlifter was disqualified for using a diuretic, two more forfeited bronze metals after testing positive for steroids, and a visually impaired tandem cyclist lost his silver medal after his "pilot" tested positive for an anti-inflammatory drug believed to increase endurance.[23]

In their quest for an edge, athletes already are turning to genetic technology. Within 10 years of the invention of recombinant DNA engineering, in which an organism is given DNA from different sources, biotech companies were adding human genes that instruct cells to manufacture the human hormone erythropoietin (EPO) to hamster ovarian cells, causing the hamster cells to start producing the human hormone. The U.S. Food and Drug Administration (FDA) approved the resulting synthetic EPO to treat anemias, illnesses triggered by chemotherapy and other conditions that leave patients with too few red blood cells, which EPO treats by stimulating the bone marrow to produce more red blood cells. It didn't take long for athletes to realize that the hormone also could enhance performance, since the additional red blood cells would deliver more oxygen to the muscles, enabling them to produce more energy. As a result, the use of synthetic EPO was banned in elite sports in the 1990s.[24]

Two other human hormones that may enhance athletic performance, human growth hormone and insulin-like growth factor 1 (IGF-1), also have been manufactured using recombinant DNA. Similar to Gregory Stock's poison gene that only produces its cancer-killing chemical in the presence of an injected hormonal trigger, there is even a "gene factory" called Repoxygen that, when inserted into muscle cells, produces EPO only when it senses low amounts of oxygen in the surrounding tissue, such as when oxygen is depleted during exercise. This worries anti-doping officials because the additional amounts of EPO intermittently produced by Repoxygen may not be detected by drug tests. Athletes are also interested in a protein called myostatin; the less of it you have, the bigger your muscles. Techniques to reduce the amount of myostatin in the body are being tested to treat muscle-wasting diseases such as muscular dystrophy, and researchers have developed an antibody that blocks the production of the protein in mice.[25]

Performance enhancement by directly modifying DNA is a more distant prospect for humans, but researchers led by Ron Evans at the Salk Institute have modified a gene in mice to increase the production of a protein called PPARdelta. The original idea was to produce leaner mice because PPARdelta improves their ability to break down fat. The genetic modification that Evans and his colleagues used turned out to have an added bonus, however: the extra protein also increased the number of "slow-twitch" muscles in the mice, the type of muscles that endurance athletes such as marathon runners develop. Sure enough, Evans's mice behaved like they had been training to run marathons. Whereas normal mice placed on treadmills can go about 900 meters before tiring out, the mice with the modified gene ran twice that far.[26] What's more, the offspring of these supermice inherited the same stamina-increasing gene, making this a successful example of germ line genetic enhancement.

Eventually, someone is bound to try to make these sorts of genetic alterations in primates, followed by humans. The first customers may be athletes looking for a performance advantage. Improvements in performance also are of interest to the military. A chapter entitled "Be All You Can Be" in Joel Garreau's 2005 book *Radical Evolution: The Promise and Peril of Enhancing Our Minds, Our Bodies—and What It Means to Be Human* discusses how genetic engineering could be used to make supersoldiers. If whales and dolphins slept the way humans do, for example, they'd drown, so half of their brains stay awake while the other half sleeps. If we could figure out how to engineer this in humans, Garreau claims, soldiers could remain partially alert while they slept. Garreau also suggests that, since the mitochondria in the cell provide the energy that enables the cell to function, multiplying the number of mitochondria per cell ought to give soldiers more energy.[27]

Some of the ideas put forward by transhumanists are more fanciful. Why not give people four-color vision, asks Lee Silver. Silver predicts that being able to perceive a four-color world, rather than the normal three colors of red, green, and blue, would produce a sensory experience similar to that of a color-blind person, who can see in only one or two colors, suddenly being able to see all three. Silver also imagines giving people a sense organ that can "see" radio frequencies and is then able to both send and receive radio waves, so that people can transmit their thoughts radiotelepathically.[28] Garreau agrees that humans will become telepathic and speculates that our minds will become able to heal wounds by radioing repair instructions to the affected tissues.[29]

Transhumanists also have few qualms about adding animal traits to humans. Oxford professor Julian Savulescu sees no reason, for example, why people couldn't develop the visual acuity of hawks.[30] As Silver observes, "if something has evolved elsewhere, then it is possible for us to determine its genetic basis and transfer it into the human genome."[31] Humans could easily acquire "relatively simple animal attributes," such as ultraviolet or infrared vision, which would enable us to see much better at night, along with light-emitting organs from fireflies and fish, electric generators from eels, and magnetic metal detectors and compasses from birds. More advanced animal capabilities would likely follow, such as a dog's sense of smell, which allows it to detect and identify molecules present in the air in tiny concentrations, and a bat's sonar system, which would permit people to move about securely in total darkness.[32]

Although increasing technical difficulty is undoubtedly a main reason why transhumanists contemplate a gradual progression for human genetic engineering, they clearly are sensitive to the need to move cautiously in order to prevent a backlash that would threaten their agenda. Hence, it comes as no surprise that they reserve the transformation of the mind, which we associate most closely with our sense of personhood and self-identity, for last.

High on the list of mental alterations is improving the way humans think. Transhumanists expect that genetic modifications will make us more intelligent, give us better memories, and ramp up our powers of concentration. Simon Young predicts that humans will reason, learn, and create at much higher speeds.[33] Joel Garreau wants to add photographic memories, total recall, and the ability to read books in minutes.[34] James Hughes comments that while "pessimists insist that intelligence is far too complex to be changed with gene therapy, it appears we will be able to tweak our own intelligence with gene therapies targeted at 'g genes,' and increase the intelligence of our children through germinal choice."[35] (G genes code for proteins that affect the speed at which information passes in and out of cells.)[36]

Researchers at the University of Pennsylvania, for example, have created a strain of "smart" mice that produce more of a protein called NR2B, giving them superior memory and learning abilities.[37] Studies in humans have identified a gene called dysbendin on chromosome 6 and another called SNAP-25 on chromosome 20 that are associated with cognitive ability.[38] 23andMe, the company mentioned in the previous chapter which furnishes genetic testing directly to consumers over the Internet, sells a test for the SNAP-25 gene that the company claims can provide information about "measures of intelligence." Canadian researchers have identified a gene that codes for a protein called eIF2Ð that impairs memory, and they have shown that mice with a genetic mutation that prevents the protein from being produced perform better on memory tests.[39] A study funded by the NIH found that people with a certain variant of the gene catecholamine-O-methyltransferase (COMT) can improve both memory and problem solving by taking a drug called tolcapone, which is prescribed for patients with Parkinson's disease.[40] A Chinese research institute located in Hong Kong has purchased more than $90 million worth of advanced genetic sequencing machines and is using them, among other things, to sequence the DNA of 2,000 Chinese schoolchildren to search for genes that correlate with educational testing scores.[41]

Transhumanists want to go beyond improving memory and cognition, however. Silver proposes engineering the mind to make people better at business, art, and music. Stock, noting that "gene therapy has been used to turn lazy monkeys into workaholics by altering the reward centre in the brain,"[42] suggests that we can enhance industriousness, and he also speculates that perfect pitch may be just a matter of altering a single gene.[43] Garreau says we'll do away with the need to sleep.[44]

Transhumanists also recognize the danger that if people become cleverer and stronger, they might use their newfound abilities for ill-gotten gain. Economist Robin Hanson, for example, imagines an enhancement that would make it easier to deceive people: "One tempting enhancement will be to better control the various subconscious clues we give off, so that we can lie more convincingly. We might put our facial expressions under more conscious control, for example, and use hidden internal speed improvements to help us fake reaction times."[45] Transhumanists therefore recommend that evolutionary engineering also eradicate antisocial behaviors such as criminality and extreme aggression.[46]

While parents already use relatively crude techniques such as sperm sorting, IVF, and abortion following prenatal testing to avoid giving birth to a baby of the wrong gender, in the future, gender likely will be determined directly by genetic manipulation. For that matter, says Stock, genetic engineers could establish the person's sexual orientation,[47] and Australian philosopher Peter Singer points to genetic experiments on promiscuity in which "researchers used gene therapy to introduce a gene [the vasopressin receptor gene] from the monogamous male prairie vole, a rodent which forms life-long bonds with one mate, into the brain of the closely related but polygamous meadow vole. Genetically modified meadow voles became monogamous, behaving like prairie voles." This form of genetic alteration, notes Singer, should appeal particularly to Christians.[48]

In addition, evolutionary engineering might be employed simply for fun. No need to put up with such dull-colored skins, Naam observes, when installing genes from tropical birds and fish that code for protein dyes would give us skin and hair in vibrant red, blue, and yellow hues. Or we could import animal genes that make us glow fluorescent green in the dark,[49] like the marmosets mentioned in the introduction.

Perhaps in keeping with the current public health campaign against obesity, transhumanists also extol modifying genes to make us svelte. Futurists Ray Kurzweil and Ramez Naam suggest that genetic engineering could enable people to eat as much as they wanted while remaining thin.[50] Kurzweil describes a conversation he had with James Watson, the codiscoverer of the double-helix structure of DNA, in which Watson predicted that in 50 years there would be drugs that would allow people to eat without gaining weight. "Fifty years?" scoffed Kurzweil. "We have accomplished this already in mice by blocking the fat insulin receptor gene that controls the storage of fat in the fat cells. Drugs for human use . . . will be available in five to ten years, not fifty."[51] The problem, explains James Hughes, is that we still have the metabolism that we first developed when we became modern humans: "The basic cause of obesity is that we have bodies designed to spend hours walking around the savannah every day and brains that find easy access to fats, sugar, and carbohydrates irresistible."[52] Genetic engineering will allow us to correct this evolutionary time lag, Hughes believes. He notes that "more than sixty pharmaceutical treatments to alter metabolism or reduce appetite are being developed, based on more than 130 genes that have been discovered to regulate weight in humans. The gene or drug tweaks that keep us slim will likely be much simpler, modifying just one of those chemical pathways."[53]

For Hughes, the key to happiness for our descendants will not just be the ability to be gluttons without gaining weight, however. The fact that sunny dispositions seem to travel in families, he says, indicates that we can use drugs and genetic modification to "jack our happiness set point to its maximum without negative side effects."[54] Naam adds that genetic modification will make us friendly, romantic, daring, and empathetic.[55] Joel Garreau interviewed a philosopher named Gregory Pence who predicted that our descendants will be engineered to have a more acute sense of wonder and curiosity and to tell better jokes.[56] And he repeats Nick Bostrom's observation that evolutionary engineering could open us up, in Garreau's words, to "pleasures whose blissfulness vastly exceeds what any human has yet experienced." According to Garreau, Bostrom

can imagine much cleverer philosophers than us. He can imagine new and different kinds of artworks being created that would strike us as fantastic masterpieces. He can imagine a love that is stronger and purer than any of us has ever felt—including preserving romantic attachment to one's partner undiminished by time. Our thinking about what is possible for humans to attain is likely constrained by our narrow experience, [Bostrom] believes. We should leave room for the possibility that as we develop greater capacities, we will discover values that will strike us as more profound than those we can realize now, including higher levels of moral excellence.[57]

But future humans won't only be healthy, strong, brilliant, blissful, fluorescent, and more ethical. They will live longer, indeed much longer, and may even attain transhumanism's ultimate goal, to live forever. "The Holy Grail of enhancement," says John Harris, "is immortality."[58]

The desire to live forever is nothing new, of course. The Babylonian *Epic of Gilgamesh*, perhaps the oldest known literary work, tells of a king who searched for the key to immortality. The medieval alchemists sought not only to transmute base metals into gold but to discover the "philosopher's stone" that would prevent aging. In the early twentieth century, men slept with virgins and ate ground-up, hormone-containing monkey testicles to keep themselves young. So far as is known, no one appears to eat monkey testicles anymore to prevent aging, but Americans spend more than $45 billion a year on anti-aging products and services, most of which are probably ineffective, and the American Academy of Anti-Aging Medicine, founded by osteopaths Ronald Klatz and Robert Goldman in 1993, now boasts almost 20,000 physician members.[59]

One popular potentially anti-aging product is human growth hormone (HGH), which used to be available only in small quantities when it was extracted from the pituitary glands of corpses. But since 1985 it has been synthesized in virtually unlimited quantities using recombinant DNA technology, and as many as 30,000 Americans pay in the neighborhood of $1,000 a month to have it injected for anti-aging purposes.[60] This is pretty amazing not only because there is no scientific evidence that it works, but also because prescribing HGH to prevent aging is a federal felony.[61] (This is due to a quirk in the federal food and drug laws that Congress put there as part of its effort to combat the use of performance-enhancing drugs in sports. In contrast to all other drugs, which doctors may prescribe for whatever medical purpose they wish, a practice known as "off-label use," HGH may only be prescribed for the specific uses that the FDA has approved, and fighting aging isn't one of them.) Leonard Hayflick, a leading gerontologist, scoffs at the gullible consumers who buy HGH and their physicians who are all too willing to oblige. "Yesterday's prolongevists who searched for the fountain of youth, advocated sleeping with young virgins, encouraged grafting of monkey testicles, and ate yogurt," he chaffs, "simply have been replaced with modern equivalents, who have equal probability for success."[62] Not only is there little evidence of success with these approaches, but anti-aging hormone treatments have been shown to cause diabetes and glucose intolerance[63] and increase the risk of cancer,[64] dementia,[65] coronary heart disease, stroke, and pulmonary embolism.[66]

While effective life-extending treatments are not yet available, serious scientific work is underway. Spurred on by the NIH, whose Institute on Aging in 2001 declared "Unlocking the Secrets of Aging, Health and Longevity" to be one of its strategic goals, researchers are pursuing an approach called "caloric restriction," which entails consuming only 70 percent or less of the normal daily caloric intake. Roundworms, fruit flies, and mice on such diets live an average of 40 percent longer. Since sticking to such a regimen is likely to be unrealistic for most people, the search is on for drugs that will produce the same result. One substance of interest is resveratrol, which comes from grape skins and may be a major explanation for why French people, who eat so much rich food, have not died out from heart disease.[67]

The transhumanists' campaign against death is likely to follow much the same pattern as the other parts of their agenda. John Harris maintains that "we do not die of old age but of the diseases of old age. It is species typical of us to die of these as we normally do, but it is not necessarily necessary that we do."[68] Initial efforts to prevent aging therefore will wield genetic engineering as a weapon against the illnesses of old age and will be passed off as an unremarkable use of gene therapy in the overall battle against disease. More direct attempts to slow or stop the aging process will follow. The first step will be to identify genetic causes of aging and disable them. According to Simon Young, "we have learned to identify the genetic recipe for life. There is no reason to suppose that we will not go on to identify the genetic program for death."[69] Young cites the prediction by controversial anti-aging enthusiast Aubrey de Grey that humans soon will be living until they are 1,000 or even 5,000 years old. "Breakthroughs have been made," Young proclaims. "The mechanisms by which cells wear out, and the chemicals which extend their life, have been identified."[70]

Young is referring here specifically to telomeres, repetitive sequences of DNA that dangle from the ends of chromosomes. Each time a cell divides, some of this DNA is lost. When enough of it disappears, the cell stops dividing and reaches the end of its lifetime, and when enough cells stop dividing, the body, no longer able to repair itself, perishes. But in 1984, researchers discovered an enzyme called telomerase that bears the instructions for replacing the lost DNA. Anti-aging enthusiasts immediately seized upon telomerase as a key clue to living longer by preventing cells from dying. In 2009, the discoverers of telomerase were awarded the Nobel Prize in Physiology or Medicine.

One small problem arises, though. If telomerase prevents cells from dying, they go on to divide indefinitely and indeed become immortal. There is a name for immortal cells, however: cancer. In fact, the life span of our cells appears to be a balance between longevity and disease, with death the price we pay for avoiding malignancy. As disability rights advocate Anita Silvers explains, the gene which helps regulate cell-division raises its production as individuals age, increasing about ten-fold from age 20 to age 70. Because of this gene, as we grow older, the process whereby adult stem cells renew tissues slows down radically. But while this halting produces the symptoms of ageing it can also be beneficial because it helps to inhibit the growth of cancer. The gene offsets the increasing risk of tumors spreading by gradually reducing the ability of cells to proliferate. In teleological terms, we might say that nature makes our body choose why we will die, trading increased vulnerability to deterioration from old age for an increased defense against death from the cumulative cell damage that ends up with cancer. The same biological process responsible for wrinkles, grey hair and weak bones is the one that fights cancer.[71]

Another problem with telomerase is that telomere shortening apparently does not work the same way in other animals as it does in humans and other primates; consequently, there is no known animal model in which the action of the enzyme can be tested before we try it on ourselves.

The risk that we might invite a premature death from cancer by prolonging the process of cell division has dampened most of the interest in telomerase, but not all. One company in New York sells a dietary supplement made from an extract of a Chinese herb called astragalus, which it claims triggers the release of the enzyme. A year's worth of the product costs $14,000.[72] The company runs a website for its adherents, whom it calls "Telonauts." According to the website, these are people who "refuse to sit by idly as time takes its toll on our bodies and minds when there is solid scientific technology (telomerase activation and other cool stuff) available to prevent this."[73]

It is unclear what "other cool stuff" the website is referring to, but it might be cloned and cultured stem cells, which James Hughes regards as a potential means of replacing worn-out body parts.[74] "[W]hat seniors diagnosed with dementia—and even people over 20 years old— really need," Hughes says, "is a way of reversing brain damage. Stem cell research has shown that the brain is able to repair itself by growing new neural stem cells well into the senior years. As research unravels how 'neurotrophic' chemicals govern the growth of neural stem cells in the brain, we will begin to develop drugs that encourage brain self-repair. . . . Even better than a neurotrophic pill would be a neurotrophic gene therapy that helped the brain to repair itself."[75] According to Hughes, genes introduced into the brains of Alzheimer's sufferers have been shown to stimulate the production of nerve growth factor in the brain, helping to reverse atrophy.[76]

Other "cool stuff" to combat aging might be genes from turtles, rockfish, and other animals that live long lives, which Julian Savulescu thinks hold the promise of longevity.[77] Or it might be the work of a group of Cambridge, Massachusetts, researchers that includes a Boston University Medical School professor named Thomas Perls. In 1999, Perls started a project called the New England Centenarian Study, which by 2006 had enrolled more than 800 people who were at least 100 years old in order to investigate their genes. By 2001, Perls and his colleagues were focusing their interest on chromosome 4, and in 2003, they announced that they had pinpointed an association between longevity and a gene on chromosome 4 called MTP, which plays a role in cholesterol metabolism.[78] In 2005, a group of German scientists declared that they could not repeat Perls's results in larger populations of older people in France and Germany.[79] Four years later, the German group identified a different genetic variation, FOXO3A, as being associated with longevity in their German subjects.[80] Perls's team continues to examine other genes, including one called CGX1 and another they have named INDY, for "I'm Not Dead Yet," after a line in the film *Monty Python and the Holy Grail*.[81]

Transhumanists find little point in living forever in old bodies, however, even in bodies that remain healthy. So in addition to being immortal, they want humans to engineer themselves to be forever young. Ray Kurzweil, for example, is counting on cloning and stem cells to do the trick, the same technologies that John Harris wants to employ to eliminate the diseases of old age. Our bodies will be rejuvenated, says Kurzweil, "by transforming your skin cells into youthful versions of every other cell type."[82]

꒯

Eternal life, perpetual happiness, and permanent youthfulness certainly seem worth looking forward to, and therefore it is not surprising that transhumanists articulate them as the goals attainable by controlling future human evolution. What is striking, however, is how closely the transhumanist vision resembles the images of heaven at the heart of most of the world's religions. The main difference is that, with the exception of those who arrange to have their bodies or at least their heads cryogenically preserved when they die, today's transhumanists cannot hope to enter into their paradise. Yet they regard this misfortune as all the more reason to spare their descendants the same fate. Transhumanism, in short, bears a striking resemblance to religion. It seeks to provide hope in the face of death, a measure of control over the savage aspects of nature, and meaning to its followers' existence. No wonder that there is a Mormon Transhumanist Association according to whose creed transhumanism is a means of realizing "diverse prophetic visions of transfiguration, immortality, resurrection, renewal of this world, and the discovery and creation of worlds without end."[83]

The transhumanists may yearn for a future of eternal human happiness, but they aren't very happy right now with one another. They once belonged to a single organization, the World Transhumanist Association (WTA), but they have now split into two factions. Hughes heads one, called Humanity+. Hughes is convinced that transhumanism must foster liberal democracy. He calls his movement "democratic transhumanism" and explains that it advocates "both the right to use technology to transcend the limitations of the human body and the extension of democratic concerns beyond formal legal equality and liberty, into economic and cultural liberty and equality, in order to protect values such as equality, solidarity, and democratic participation in a transhuman context."[84] The other faction, led by Simon Young, comprises libertarian transhumanists. "I've founded the World Transhumanist Society to counter the dominance of the 'World Transhumanist Association' by Marxists," blogs Simon. "If you reject socialism (state control of resources), you really should reject the WTA: it's run by a Marxist who advocates a world state that distributes eugenic enhancements as it sees fit . . . the ultimate nightmare of a Brave New World."[85]

Young and Hughes may disagree about the politics of transhumanism, but they have no doubts about the value of the technology or its ability to improve human existence. Fictional depictions of the future of evolutionary engineering, however, are far from optimistic, with tech technology at best a mixed bag and at worst the destroyer of human society. Consider how far from perfect, for example, are the imagined lives of genetically engineered superheroes. In comic books, Captain America feels excessively guilty about the death of his sidekick, the Hulk has a dangerously short fuse, and the X-Men are scorned as "mutants." In the movie *Soldier*, the hero played by Kurt Russell is a highly trained and carefully conditioned Special Forces soldier who is replaced and eventually hunted down by soldiers who have been genetically engineered to be superior, yet Russell whips them all single-handedly. In the Star Trek movie *The Wrath of Khan*, Khan is a genetically engineered supertyrant. The main character in the television show *Dark Angel*, played by Jessica Alba, is a genetically engineered soldier whose design defect, an inability to manufacture serotonin, makes her suffer seizures. In his 1966 novel *The Eyes of Heisenberg*, Heinlein describes a distant future in which Earth is ruled by a small number of genetically engineered "Optimen" who rigidly control the genetic makeup and reproductive behavior of the rest of the population, which also has been genetically manipulated, but not to the extent of the Optimen. While everyone lives long lives by virtue of taking special enzymes, the enzymes make the Optimen immortal. Yet the Optimen live in a culturally sterile and highly artificial environment and eventually prove unable to maintain their fragile hormonal balance.[86]

A recurrent theme in fictional accounts of human genetic engineering is the exploitation of persons who are genetically modified. In the TV series *Mutant X*, an evil corporation tries to take advantage of mutated offspring whom it secretly engineered. Ridley Scott's classic film *Blade Runner* features humanoids called "replicants" which the powerful Tyrell Corporation created from DNA to perform military and other tasks on distant planets and, in the case of the "pleasure model," to serve as sex slaves. The replicants are extraordinarily strong and agile but have crimped emotions, and several of them return to Earth to try to extend their brief life spans. In Lois McMaster Bujold's Nebula Award–winning 1988 science fiction novel *Free Falling*, a mining company sends Leo Graf, one of its engineers, to a distant planet to serve as a welding instructor. His students turn out to be youngsters whom the company has genetically engineered to be well suited for working in zero-gravity environments. The first one Graf meets is Tony, who is told to go over and shake hands. "Tony pulled himself over obediently to the control panel. He wore red shorts—Leo blinked, and caught his breath in shock. The boy had no legs. Emerging from his shorts were a second set of arms."[87] These youngsters, called "quaddies," are treated by the mining company as its property ("post-fetal experimental tissue cultures") rather than as persons, since the company "made" them.

In these fictional futures, exploitation goes hand in hand with corporate greed. People are being engineered primarily for profit, rather than for their own good. Tyrell, the man whose company by the same name creates the replicants in *Blade Runner*, boasts that "commerce is our goal here at Tyrell. More human than human is our motto."[88] Michael Crichton's 2006 novel *Next* opens with a biotech venture capitalist proclaiming to a conference of fellow genetic entrepreneurs that "progress is our mission, our sacred calling. Progress to vanquish disease! Progress to halt aging, banish dementia, extend life! A life free of disease, decay, pain, fear! The great dream of humanity—made real at last!" Then comes the payoff: "Sixty billion this year. Two hundred billion next year. Five hundred billion in five years! That is the future of our industry, and that is the prospect we bring to all mankind!"[89] Like the transhumanists concerned about the reactions of their opponents, the venture capitalist focuses on the health benefits of genetic engineering. "Of course, we face obstacles to our progress," he acknowledges. "Some people—however well intentioned they *think* they are—choose to stand in the way of human betterment. They don't want the paralyzed to walk, the cancer patient to thrive, the sick child to live and play. These people have their reasons for objecting, Religious, ethical, or even 'practical.' But whatever their reasons, they are on the side of death. And they will not triumph!"[90]

Another common theme in fictional versions of evolutionary engineering is that engineered people will face discrimination from normal people. Robert Heinlein's 1982 novel *Friday*, for example, describes the plight of "artificial humans," genetically enhanced beings who are not recognized as persons under the law, deemed by religions to be without souls, and prohibited from marrying or owning property.[91] In Nancy Kress's 1993 *Beggars in Spain*, a small number of children are genetically engineered to live without needing to sleep. Known as "the Sleepless," they are "more intelligent, better at problem-solving, and more joyous."[92] Their greater intelligence and problem-solving ability result from being able to conserve the expenditure of the energy necessary to make REM sleep possible. The lack of REM sleep also accounts for the improvement in their mood; researchers discover that eliminating REM sleep also prevents depression. When the first Sleepless person dies in an auto accident at the age of 17, moreover, the autopsy reveals that, owing to a healthier immune system, his tissues have been regenerating themselves. In other words, the Sleepless do not age.

The reaction of the rest of the population, called "the Sleepers," is one of jealousy and fear, and one of the Sleepless is even murdered. The rancor they face perplexes one Sleepless person. "There have always been haters . . . , hate Jews, hate Blacks, hate immigrants," she observes. "It doesn't mean any basic kind of schism between the Sleepless and the Sleepers."[93] But another Sleepless corrects her, telling her that she is

a different kind of person entirely. More evolutionarily fit, not only to survive but to prevail. Those other objects of hatred you cite—they were all powerless in their societies. They all occupied *inferior* positions. You on the other hand—all three Sleepless in Harvard Law are on the *Law Review*. . . . Every Sleepless is making superb grades, none have psychological problems, all are healthy, and most of you aren't even adults yet. How much hatred do you think you're going to encounter once you hit the high-stakes world of finance and business and scarce endowed chairs and national politics?[94]

The discovery that the Sleepless do not age exacerbates their social predicament. "Tampering with the law of nature has only brought among us unnatural so-called people who will live with the unfair advantage of time," says one anti-Sleepless news source. "Time to accumulate more kin, more power, more property than the rest of us will ever know," warns another. "How soon before the Super-race takes over?" clamors a third.[95] Then comes the additional revelation that sleeplessness, in addition to being engineered into the germ line, is a dominant trait; not only will the next generation of babies born to Sleepless parents be Sleepless, but so will any offspring of a Sleepless parent who reproduces with a Sleeper.[96] This only adds fuel to the fire.

The Sleepers react with intolerance. Sleepless are prevented from renting apartments because "their prolonged wakefulness would increase both wear-and-tear on the landlord's property and utility bills." They are not allowed to operate 24-hour businesses because they have an unfair competitive advantage. They are banned from jury duty because a jury that they served on would not be a jury of "one's peers." Georgia declares sex between a Sleepless and a Sleeper to be a third-degree felony, the same as bestiality.[97] Although Sleepers avail themselves of a number of types of genetic engineering, including "enhanced IQ, sharpened sight, a strong immune system, [and] high cheekbones," they do not opt for Sleeplessness. "Genetic alterations were expensive," writes Kress. "Why purchase for your beloved baby a lifetime of bigotry, prejudice and physical danger? Better to choose an assimilated genemod. Beautiful or brainy children might encounter natural envy, but usually not virulent hatred. They were not viewed as a different race, one endlessly conspiring at power, endlessly controlling behind the scenes, endlessly feared and scorned. The Sleepless . . . were to the twenty-first century what Jews had been to the fourteenth."[98]

Eventually the Sleepless take refuge in a 150-square-mile fortified zone called the Sanctuary, and finally they are hounded off the planet to an orbiting space station that they have constructed.[99] Society ultimately divides into three classes: the small number of Sleepless who provide the technical, scientific, and financial know-how to run the automated production systems; a larger group of "donkeys" who are genetically enhanced though not Sleepless and comprise the politicians and bureaucrats who run the government and the economy; and the "Livers," who make up 80 percent of the population and, receiving everything they need for free, live a life of leisure. Ultimately the Sleepless decide that they need children more advanced than themselves to perpetuate their domination of the world and to avoid "the well-known phenomenon of intellectual regression to the mean, in which superior parents have children of only normal intelligence,"[100] so they engineer a new, more advanced human called Superbrights.

In the 1997 film *Gattaca*, the discrimination runs in the other direction. The future society the film depicts is ruled by those with superior genes. Parents use PGD to select the two or three best embryos to implant in the womb from the dozen or so that have been fertilized in vitro, and they also appear to manipulate the embryos' genes. When the parents of the hero, Vincent, decide to have another child, for example, the geneticist they consult tells them that he has "taken the liberty of eradicating any potentially prejudicial conditions—premature baldness, myopia, alcoholism and addictive susceptibility, propensity for violence, obesity." He also assuages their concerns about whether they should leave some things to chance by saying, "You want to give your child the best possible start. Believe me, we have enough imperfections built in already. The child doesn't need any additional burdens. Keep in mind this child is still you, simply the best of you. You could conceive naturally a thousand times and never get such a result." Vincent, who is an "in-valid" because he was conceived without genetic enhancement, fakes his way into the privileged class of "valids" by using an engineered person's DNA to fool the ubiquitous DNA analyzers. "I belong to a new underclass, no longer determined by social status or the colour of your skin," Vincent explains. "We now have discrimination down to a science." Indeed, when he applies for a job at Gattaca Aerospace Corporation, where he eventually hopes to become an astronaut, Vincent's job interview consists entirely of a urinalysis to test his genes, which he passes by substituting a sample of the engineered person's urine.

When people think of the ultimate scientific dystopia in which genetically superior humans rule the roost, what often comes to mind is Aldous Huxley's *Brave New World*. In this novel, however, humanity is enslaved by combining a reproductive cloning method called the "Bokanovsky process" which enables large numbers of identical members of the four lower human castes to be produced from a single egg with extensive behavioral conditioning and the use of the psycho-pharmaceutical soma, which Huxley describes as having "all the advantages of Christianity and alcohol, none of their defects."[101] (In *Brave New World*, soma is a psychoactive substance. There actually is a drug called "Soma," a brand name for the muscle relaxant carisoprodol.) As unappealing as it is, Huxley's future society is not the product of genetic engineering, probably because the idea was too far advanced for his time.

The most disturbing depiction of human genetic engineering is in fact Margaret Atwood's *Oryx and Crake*. The book jumps back and forth through the life of a character named Jimmy (the Snowman), but chronologically, the story begins with society having divided into the privileged class, who work for corporations and live in gated compounds, and the hoi polloi, who live in the dirty, unruly "pleeblands." Some of the corporations pursue cosmetic biotechnology. One of their goals, for example, "was to find a method of replacing the older epidermis with a fresh one, not a laser-thinned or dermabraded short-term resurfacing but a genuine start-over skin that would be wrinkle- and blemish-free." After all, "what well-to-do and once-young, once beautiful woman or man, cranked up on hormonal supplements and shot full of vitamins but hampered by the unforgiving mirror, wouldn't sell their house, their gated retirement villas, their kids, and their soul to get a second kick at the sexual can? Nooskins for Olds, said the snappy logo."[102]

Jimmy's father is a scientific researcher for one of these corporations, and his mother eventually becomes so disgusted with her husband's projects, which include growing human brain tissue in genetically engineered animal hybrids, that she abandons the family. "You hype your wares," she complains to Jimmy's father, "and take all their money and then they run out of cash, and it's no more treatments for them. They can rot as far as you and your pals are concerned."[103] The corporations that concentrate on health care products realize that they can't survive if people don't get sick, so in addition to selling cures and genetic preventives, they insert genetically engineered pathogens into their premium brands of vitamins. "Naturally they've developed the antidotes at the same time as they're customizing the bugs," Jimmy's brilliant but deranged idol Crake tells him, "but they hold those in reserve, they practice the economics of scarcity, so they're guaranteed high profits."[104]

In Atwood's novel, animal genetic engineering has become bizarre. Jimmy encounters "a large bulblike object that seemed to be covered with stippled whitish-yellow skin. Out of it came twenty thick fleshy tubes, and at the end of each tube another bulb was growing. . . . 'Those are chickens,' said Crake. 'Chicken parts. Just the breasts, this one. They've got ones that specialize in drumsticks too, twelve to a growth unit.'" Jimmy says that he can't see any heads. "That's the head in the middle," explains Crake. "There's a mouth opening at the top, they dump the nutrients in there. No eyes or beak or anything, they don't need those."[105] The corporation is thinking of calling them "ChickeNobs," adds Crake. "They've already got the takeout franchise operation in place."[106]

When Jimmy reaches adulthood, Crake wipes out virtually the entire population with a virus he creates so that he can repopulate the planet with a new type of humanoid that he has genetically engineered called the Children of Crake. As Crake explains, "once the proteonome had been fully analyzed and interspecies gene and part-gene splicing were thoroughly underway, [creating the Children] or something like it had been only a matter of time."[107] Crake has made them "amazingly attractive . . . each one naked, each one perfect, each one a different skin colour—chocolate, rose, tea, butter, cream, honey—but each with green eyes."[108] Moreover, he has painstakingly designed the Children to be well adjusted. "What had been altered," Jimmy learns from Crake,

> was nothing less than the ancient primate brain. Gone were its destruc-
> tive features, the features responsible for the world's current illnesses.
> For instance, racism . . . had been eliminated in the model group,
> merely by switching the bonding mechanism: the . . . people simply
> did not register skin colour. Hierarchy could not exist among them,
> because they lacked the neural complexes that would have created it.
> Since they were neither hunters nor agriculturalists hungry for land,
> there was no territoriality. . . . They ate nothing but leaves and grass
> and roots and a berry or two; thus their foods were plentiful and
> always available. . . . In fact, as there would never be anything for these

people to inherit, there would be no family trees, no marriages, and no divorces. They were perfectly adjusted to their habitat, so they would never have to create houses or tools or weapons, or, for that matter, clothing. They would have no need to invent any harmful symbolisms, such as kingdoms, icons, gods, or money.

Jimmy interrupts Crake at this point. "Excuse me," he says. "But a lot of this stuff isn't what the average parent is looking for in a baby. Didn't you get a bit carried away?" "I told you," Crake replies. "These are the floor models. . . . We can list the individual features for prospective buyers, then we can customize. Not everyone will want all the bells and whistles, we know that. Though you'd be surprised how many people would like a very beautiful, smart baby that eats nothing but grass. The vegans are highly interested in that little item. We've done our market research."[109]

One of Crake's more exotic alterations is the Children's method of reproduction: the women go into heat every three years. (Interestingly, Jessica Alba's genetically engineered character in *Dark Angel* also goes into heat, having been given feline DNA, but it happens to her three times a year.) The fact that a woman is in heat, Jimmy observes, "will be obvious to all from the bright-blue colour of her buttocks and abdomen—a trick of variable pigmentation filched from the baboons, with a contribution from the expandable chromospheres of the octopus."[110] This is for practical rather than artistic reasons. "Since it's only the blue tissue and the pheromones released by it that stimulates the males," explains Jimmy,

> there's no more unrequited love these days, no more thwarted lust; no more shadow between the desire and the act. Courtship begins at the first whiff, the first faint blush of azure, with the males presenting flowers to the females—just as male penguins present round stones, said Crake, or as the male silverfish presents a sperm packet. At the same time, they engage in musical outbursts, like songbirds. Their penises turn bright blue to match the blue abdomens of the females, and they do a sort of blue-dick dance number, erect members waving to and fro in unison, in time to the foot movements and the singing; a feature suggested to Crake by the sexual semaphoring of crabs. From amongst the floral tributes the female chooses four flowers, and the sexual ardour of the unsuccessful candidates dissipates immediately, with no hard feelings left. Then, when the blue of her abdomen has reached its deepest shade, the female and her quartet find a secluded spot and go at it until the woman becomes pregnant and her blue coloring fades.

And that is that. . . . No more prostitution, no sexual abuse of children, no haggling over the price, no pimps, no sex slaves. No more rape.[111]

The Children also were designed to drop dead when they reached 30, "suddenly, without getting sick. No old age, none of those anxieties. They'll just keel over."[112] Moreover, they have no sense of humor. "For jokes you need a certain edge, a sense of malice," Crake tells Jimmy. "It took a lot of trial and error and we're still testing, but I think we've managed to do away with jokes."[113]

There is one final take on evolutionary engineering that fiction writers share: evolutionary engineers make plenty of mistakes and their creations suffer lots of unexpected, adverse effects. In Kress's *Beggars in Spain*, there originally were twenty Sleepless babies, but one of them was shaken to death by its mother, who "could not bear the 24-hour crying of a baby who never sleeps."[114] The Sleepless at least look normal; the more advanced Superbright children, on the other hand, have heads that are large and misshapen and their brains fire so energetically that they twitch and stutter. In Atwood's *Oryx and Crake*, the corporations' search for a way to engineer new skins encountered setbacks: "Not that a totally effective method had been found yet: the dozen or so ravaged hopefuls who had volunteered themselves as subjects . . . had come out looking like the Mould Creature from Outer Space—uneven in tone, greenish brown, and peeling in ragged strips."[115]

One of the superabilities the Children of Crake possess is that they are able to heal wounds and broken bones by purring like a cat. But Crake is not able to install this feature without some setbacks. "One of the trial batch of kids had manifested a tendency to sprout long whiskers and scramble up the curtains; a couple of the others had vocal-expression impediments; one of them had been limited to nouns, verbs, and roaring."[116] Atwood blunts much of the bleakness of her account with this sort of drollery. For example, Crake develops an aphrodisiac pill that protects against sexually transmitted diseases, prolongs youth, and sterilizes those who take it. "They hadn't got it to work seamlessly yet, not on all fronts," Jimmy explains. "A couple of the test subjects had literally fucked themselves to death, several had assaulted old ladies and household pets, and there had been a few unfortunate cases of priapism and split dicks."[117]

But whereas Atwood is frequently humorous, there is nothing funny in Crichton's description of the difficulties encountered by genetic engineers in his novel *Next*. "The successful injection of transgenes," he writes, "was a tricky business, and required dozens, even hundreds, of attempts before it worked properly. . . . In reality, the task of injecting a gene into an animal and making it work more closely resembled debugging a computer program than it did any biological process. You had to keep fixing the errors, making adjustments, eliminating unwanted effects, until you got the thing working. And then you had to wait for downstream effects to show up, sometimes years later."[118]

So what does the future of evolutionary engineering hold? The differences between the transhumanists' encomiastic visions and the novelists' horrific images are striking. Will our descendants be graced with beatific transfiguration and immortality, or will they struggle to survive in frightful worlds where corporate greed and scientific hubris have brought forth exploitation, political oppression, and dreadfully damaged human or humanlike beings? The answer, of course, is that it is impossible to tell so long as the gap between the current level of technology and the discoveries needed to fashion our descendants remains so great. Yet the same embryonic state of scientific development that creates so much uncertainty about the future also increases the possibility that we can control it, that we can channel our evolution in a desirable direction. The question then is how to do this, how, in effect, to engineer evolutionary engineering, so that humanity does not unwittingly fashion itself a hell on earth, or produce an even worse scenario in which it has become extinct by its own hand. The writers who describe their dystopias so imaginatively give us no suggestions, so we must turn elsewhere. One place to start is with other scientific advances that raised fears of the destruction of the human race. What can we learn from the fact that we are still here?

CHAPTER TWO

Thinking about the Unthinkable

Worrying about threats to the future of humanity has become something of a cottage industry among academics. Cass Sunstein, Harvard law professor and director of the White House Office of Information and Regulatory Affairs under the Obama administration, has written two books on the subject: *Laws of Fear: Beyond the Precautionary Principle* in 2004,[1] and *Worst-Case Scenarios* in 2007.[2] Richard Posner, a former law professor at the University of Chicago who is now a judge on the U.S. Court of Appeals, tackled the subject in his 2004 book *Catastrophe: Risk and Response.*[3] A number of other scholars have written critiques of these authors and of one another.

Posner is a brilliant but controversial public intellectual who has produced an enormous amount of work from the "law and economics" perspective, a school of thought that originated to a large extent at the University of Chicago and holds that the best way to maximize well-being is to rely on the operation of the market. In his book on planetary catastrophe, Posner discusses a number of natural and man-made potential risks to the future of the species and applies his law and economics approach to determine what should be done to avoid them. One question he asks, for example, is how much to spend on prevention. The answer, Posner argues, requires the use of what he calls "inverse cost-benefit analysis."[4] Basically, this involves making three calculations: the amount that is actually being spent to avoid the calamity, the probability that the catastrophe will occur, and the losses if it were to do so. For example, Posner calculates that the planet is spending around $3.9 million a year to prevent cataclysmic collisions with large asteroids, and he cites expert estimates that the chances of this happening are about once in every 75 million years. If the result would be the destruction of all human life, and if we simply consider the 6 billion or so people now alive and ignore persons whom the asteroid strike would prevent from being born in the future, this would yield an implicit value of a single human life at

$50,000. This is far too low, says Posner, which therefore tells us that we are not spending enough to avert an asteroid collision.[5]

Posner himself does not address the dangers created by attempts to control human evolution, but what if we applied his approach? It is hard to calculate how much government money, if any, is being spent on preventing the demise of humanity through harmful genetic engineering. The NIH has an Ethical, Legal, and Social Implications (ELSI) program,[6] which studies and attempts to guide policy on emerging human genetic technologies. Let's assume that 10 percent of the $18 million annual ELSI budget is being spent on averting an evolutionary catastrophe. The NIH also has a Recombinant DNA Advisory Committee, which advises the NIH and the FDA on genetic engineering research, and it has an annual budget of about $2 million. Let's add that entire budget to the pot. In addition, let's assume that 10 percent of the FDA's $245 million annual budget for its Center for Biologics Evaluation and Research,[7] which regulates genetic products, is spent on regulating the safety of human genetic engineering. Combining these figures yields an annual expenditure of around $28.3 million by the U.S. government alone. If we assume that this is all that the world is spending, that the value of a human life is $1,000,000, and that only 6 billion lives would be lost if the human race were wiped out, then this yields a probability of extinction of about once in every 50 million years. This risk estimate might seem comforting, since *Homo sapiens* have only been around for around 200,000 years, but since no one knows the actual probability that directed evolution would end humanity, it is hard to tell from Posner's method if enough is being spent on prevention.

Some of the risks that Posner, Sunstein, and others worry about are the cumulative effects of lots of individual human behaviors, for example, global warming resulting from the overuse of fossil fuels. Other risks, such as those arising from artificial intelligence and nanotechnology, stem from the very substance of major scientific research programs. In certain respects, directed evolution resembles global warming, in that there would have to be many individual acts, affecting a large number of births, in order for there to be a significant impact on the future of humankind. At the same time, however, modern genetic engineering is being jump-started by massive government research programs, beginning with the Human Genome Project, which collectively receive more than half a billion dollars a year in federal funding. In thinking about how to respond to the dangers of evolutionary engineering, it therefore seems useful to consider what happened in the past when major scientific research programs raised fears that could spell the end of humanity.

The first time big science was considered to be such a threat was during the Manhattan Project. Not long before the first test of an atomic bomb at the Trinity site in New Mexico, Edward Teller, a researcher who would become known as "the father of the hydrogen bomb," speculated that the Trinity test might incinerate the planet by igniting the nitrogen in the atmosphere or the hydrogen in the oceans. His fellow scientists on the project dismissed Teller's fears as groundless, maintaining that "no matter how high the temperature, energy loss would exceed energy production by a reasonable factor."[8] In other words, at some point after detonation, the atomic reaction would play itself out. Nevertheless, while he waited out the final minutes before the Trinity explosion in a bunker with other scientists and government officials, physicist Enrico Fermi offered to bet on "whether or not the bomb would ignite the atmosphere, and if so, whether it would merely destroy New Mexico or destroy the world."[9] (The Nazis were also trying to produce an atomic bomb, and the possibility of an uncontrollable nuclear explosion had occurred to them as well. When Adolf Hitler was told about this, he is reported to have thought that it was "funny.")[10]

Another research effort that scientists worried might cause cataclysmic effects is atom smashing, where electromagnets propel charged particles into target material, generating more elemental particles. Particle acceleration has been around since the early 1930s, and old-fashioned TVs contain mini-accelerators in the form of their cathode-ray tubes, but physicists recently have built huge, powerful accelerators to search for exotic forms of subatomic matter. In 1975, when the Lawrence Berkeley Laboratory assembled an accelerator called the Bevalac by combining two less powerful machines, the SuperHILAC and the Bevatron, there was some concern that it could create an abnormal state of matter that could destroy the earth.[11] Joseph Kapusta, a physics graduate student at Berkeley at the time, recalls that most scientists rejected the speculation. For one thing, they pointed out, if an accelerator experiment could destroy the earth, the same thing should happen when cosmic rays bombard objects in space like the moon, but the moon is still there. Kapusta admits, however, that "no one really knew what to expect when nuclear matter was compressed to three-to-four times the density of atomic nuclei," and he adds his own tongue-in-cheek observation that if a cataclysmic accelerator disaster did in fact take place, "no physicist would be around to be blamed for it! Moreover, it guaranteed that no physicist would ever win a Nobel Prize for the discovery."[12]

Apocalyptic fears surfaced once again in 1999 when a more powerful accelerator, the Relativistic Heavy Ion Collider (RHIC), was brought on-line at the Brookhaven National Laboratory. The RHIC was designed to create "quark-gluon plasma," a state of matter "representative of the state of the universe when it was less than one microsecond [one-millionth of a second] old and temperatures were greater than 2 trillion degrees Kelvin."[13] One fear was that the RHIC might produce a microscopic black hole that would swallow the planet.[14] Another worry was reminiscent of Kurt Vonnegut's 1963 science fiction novel *Cat's Cradle*, in which a substance called Ice Nine turned all the world's water into a solid; the RHIC, it was similarly feared, could create "strangelets," dense objects consisting not only of "up" and "down" quarks, the constituents of neu-trons and protons, but of approximately equal numbers of up, down, and "strange" quarks. Most physicists believed that if these strangelets were formed, they would convert to ordinary matter within a thousand-millionth of a second.[15] But if by some chance they were stable and also negatively charged, they could "infect" the matter around them, keep growing, and, in the words of astrophysicist Martin Rees, master of Trin-ity College at Cambridge University and the United Kingdom's Astrono-mer Royal, "transform the entire planet Earth into an inert hyperdense sphere about one hundred meters across."[16] The Brookhaven National Laboratory duly commissioned an investigation, which concluded that any strange quarks that were created would have a positive rather than a negative charge and therefore would repel rather than bond with nearby matter.[17] Moreover, the report stated, the collider could not concentrate matter in a small enough volume to make it sufficiently dense to form a black hole.[18]

RHIC was followed by the Large Hadron Collider (LHC) project at CERN, the French initials for the European Council for Nuclear Re-search.[19] This 27-kilometer-long ring of superconducting magnets strad-dling the Franco-Swiss border is so powerful that it is expected to pro-duce fundamental particles under the conditions that existed when the universe was less than 1 picosecond (one-trillionth of a second) old.[20] As might be expected, doomsayers once again raised their voices, citing the risks of black holes, strangelets, and similar catastrophic events.[21] Maybe the Brookhaven safety committee was right about the RHIC, they argued, but the LHC was much more powerful, able to collide beams of heavy ions with thirty times the total energy of the RHIC.[22] Given boasts in the news media that the LHC was the "world's largest machine," "fastest race track on the planet," "emptiest space in the solar system," "hottest spot in the galaxy," and "world's most powerful supercomputer,"[23] who really knew what would happen? The LHC was even featured in the film *Angels and Demons*, based on Dan Brown's sequel to *The Da Vinci Code*. In *Angels and Demons*, the bad guys steal a canister of antimatter from CERN in order to destroy the Vatican but are thwarted with the help of Tom Hanks.[24]

Not long before the LHC was due to start up, two men filed a lawsuit in federal court in Hawaii seeking to stop the project launch on the grounds that, since the U.S. government had not prepared an environmental impact statement, federal funding violated the National Environmental Policy Act.[25] Little is known about the plaintiffs. One of them, Luis Sancho, is said to be a science writer and professor in Barcelona, and he describes his co-plaintiff, Walter L. Wagner, as a retired radiation safety officer who lives in Hawaii.[26] A *Fox News* report said that Sancho is a Spanish citizen living in Hawaii and Wagner "claims to have minored in physics at U.C. Berkeley, gone to law school, taught elementary science and worked in nuclear medicine at health facilities, but he doesn't appear to have an advanced degree in science."[27] The lawsuit didn't get very far; the judge dismissed it after finding that government funding of less than 10 percent of the $5.84 billion cost of the LHC was not "major federal action" and therefore did not trigger the requirement of an impact statement under the law.[28] But in her opinion, Judge Gillmor did acknowledge that the "plaintiff's action reflects disagreement among scientists about the possible ramifications of the operation of the Large Hadron Collider."[29]

Like their predecessors at the Brookhaven National Laboratory, the CERN leadership commissioned a safety study, and it, not surprisingly perhaps, concluded that the fears about the LHC were unfounded.[30] But less than two weeks after the machine was turned on for the first time, it malfunctioned and had to be shut down for at least two months for repairs. According to the *New York Times*, this occurred when "an electrical connection between two of the superconducting electromagnets that steer the protons suffered a so-called quench, heating up, melting and leaking helium into the collider tunnel." An account of the accident in *Science* underscores its destructiveness: "The rupture caused the 9000 amps of current flowing through the line to 'arc' to other nearby metal parts within the machine. In an instant, that lightning strike burned through the surrounding tube that kept the superconducting line bathed in frigid liquid helium. Boiling liquid and gas flooded the magnets' sealed casings, whose emergency relief valves were not designed to cope with such a deluge. As a result, a pressure wave shot through the machine, damaging 53 magnets and tearing some of the devices, which weigh up to 35 metric tons, from their moorings."

The consequences of the liquid helium leak could have been much worse. As the *Times* explained, the helium "is used to cool the magnets to superconducting temperatures of only about 3.5 degrees Fahrenheit above absolute zero," and stray heat from the loss of the helium "can cause the magnets to lose their superconductivity with *potentially disastrous consequences*" (emphasis added).[31] Presumably these consequences would be limited to the CERN equipment and facility, but in light of the worries about catastrophic risks, the use of the term "disastrous" is certainly noteworthy. Another illustration of the delicacy of the CERN machinery was a power outage in November 2009, when the LHC was crippled for three days after a bird dropped what CERN called "a bit of baguette" on part of the main power supply to the machine.[32]

The third enterprise that scientists feared might inadvertently destroy humanity actually involved genetic engineering, although of bacteria rather than people. In 1972, Paul Berg, a Stanford biochemist who in 1980 would win half of the Nobel Prize in Chemistry, published the results of experiments in which he "recombined" DNA from two different organisms.[33] As he tells it (see the clip of him at www.dnalc.org/view/15022-How-the-first-recombinant-DNA-was-created-Paul-Berg .html), a graduate student named Janet Merz had taken two different DNA molecules, each with different properties, used an enzyme to cut them into pieces, mixed the pieces, and then added another enzyme that reassembled them into "a molecule that now shared the properties of the two starting materials." The molecules that Berg was referring to were not mere chemicals: his laboratory took DNA from Escherichia coli, a common type of bacteria that lives in the human gut and manufactures Vitamin K_2, and inserted it into the DNA of SV40, a virus found in monkey kidneys. They thus created a new type of living organism, which shared DNA from the bacterium and the virus. Because the method resembled "recombination," part of the reproductive process in humans and other animals in which DNA molecules from the parents are combined and transmitted to their offspring after first being chopped up and reassembled into new molecules, the technique that Berg's laboratory pioneered came to be called "recombinant DNA."

Obviously, there is enormous potential in being able to create a new life-form that combines the attributes and abilities of other organisms. Why not create a bacterium that can completely digest and degrade oil spills, for example, by combining DNA from four different naturally occurring bacteria, each of which has the digestive equipment to only do part of the job? A microbiologist working for General Electric named Ananda Chakrabarty applied for a patent for just such a recombinant bacterium. When the Patent and Trademark Office denied the application, Chakrabarty went to court, and in a landmark decision, the U.S. Supreme Court upheld his patent by ruling for the first time that an individual has a right to patent a new form of life.[34] (Incidentally, General Electric never commercialized Chakrabarty's invention, which Chakrabarty blamed on the company's fear of the public outcry that would follow the intentional release of a man-made organism into the aquatic environment.[35] Another version of the story has it that, as the new bacterium reproduced, subsequent generations lost too much of their oil-digesting ability for the product to be commercially viable.)[36]

Recombinant DNA can provide even more direct benefits for human health. A few years after Chakrabarty created his oil-eating bacterium, a fledgling company called Genentech programmed E. coli with human DNA so the bacterium could manufacture a human protein. One year later, Genentech announced that its researchers, working with the City of Hope National Medical Center in Duarte, California, had spliced a human gene into E. coli that made it produce human insulin. Since then, the FDA and the European Medicines Agency have approved more than 151 drugs manufactured with recombinant DNA technology.[37]

Before the initial experiments on recombinant DNA were performed, however, there was a complication. Remember that Berg was about to create a new DNA molecule combining DNA from E. coli and SV40, a virus found in monkey kidneys. It turns out that the Salk polio vaccine that was developed during the 1950s was produced by culturing three types of polio virus in a culture made from monkey kidneys, and SV40 had been detected in samples of the vaccine. This made the vaccine researchers curious about the properties of SV40, and in 1961, they discovered that it caused cancer in hamsters.[38] This meant that there was a chance that it could cause cancer in humans as well. Berg was undoubtedly aware of this finding, but he nevertheless decided to work with SV40 because it was easy to use. By happenstance, one of his colleagues, a cancer researcher named Robert Pollack, got wind of Berg's experiment and became alarmed. In fact, the government had taken the threat to humans so seriously in the 1960s that it had screened the entire polio vaccine stock to weed out any vaccine that had been contaminated with SV40. Now Pollack's colleague Paul Berg was about to combine this cancer-causing virus with a bacterium, and not just any bacterium, but one that flourished in the human gut. Could the resulting virus-containing bacterium find its way into the human digestive system? Would Berg's students or lab workers become infected? Could this spread throughout the population, like outbreaks of food-borne E. coli? Would there be any way to stop it?

Pollack confronted Berg with his fears. To his credit, Berg immediately agreed to stop his experiments and relayed Pollack's warning to colleagues at the next annual Gordon Research Conference on Nucleic Acids. Along with nine other scientists, among them four Nobel laureates, Berg published a letter in the journal *Science* calling for a voluntary, worldwide moratorium on further recombinant DNA experiments "until the potential hazards of such recombinant DNA molecules have been better evaluated or until adequate methods are developed for preventing their spread."[39] Researchers around the world heeded the letter, and so far as is known, recombinant DNA experiments did not resume for over a year until the NIH established safety guidelines to govern the experiments.[40]

～

What can we learn from these episodes? One of the things that they have in common is that, in each case, the very scientists who were involved in designing and conducting the projects were relied upon to lay the safety concerns to rest. In fact, the Berg moratorium is held up as a signal event in the regulation of science, the only known instance in which researchers around the world successfully policed themselves to avert a potential catastrophe. So can we trust genetic scientists to protect us from other potentially dangerous undertakings, such as attempting to control human evolution? Richard Posner, for one, doesn't think so. "Few scientists," he argues, "have the time, the background, or the inclination to master the alien methods of public policy," and to their way of thinking, "measures of protection against dangerous knowledge, such as knowledge of how to use gene splicing to create a more lethal pathogen, are simply an impediment."[41]

Posner's answer to the danger of relying on self-interested scientists to police themselves is to delegate the regulation of science to lawyers, since "policing the intersection between law and science is a more natural role for lawyers to play."[42] He concedes, however, that few lawyers have a sophisticated understanding of science, and so they find themselves having to rely for scientific expertise on the selfsame scientists they are trying to regulate. Posner's solution is to change the way lawyers are trained. Law schools, he says, should require that "a substantial fraction of law students be able to demonstrate by the time they graduated . . . a basic competence in college-level math and statistics plus one science such as physics, chemistry, biology, computer science, medicine, public health, or geophysics."[43]

How realistic is Posner's solution? Consider the CERN report mentioned earlier on the safety of the LHC. The discussion of the risk that the collider would generate a black hole runs a little more than four pages and contains 18 mathematical formulas like this one:[44]

$$\Gamma_{\mathrm{D}} \quad \approx \quad T_{\mathrm{BH}}^{4+d}\, R_{\mathrm{S}}^{2+d}$$

$$= \quad M_*^2 \left(\frac{M_*}{M}\right)^{\frac{2}{2+d}}$$

According to the chair of the Physics Department at a major research university, given the technical level of the discussion, it would take a minimum of advanced graduate work in theoretical physics to be able to understand and judge the persuasiveness of the findings in the CERN report (Daniel Akerib, Case Western Reserve University, Feb. 17, 2009). Lawyers and other lay regulators are going to have to continue to rely on scientific experts after all to help police scientific research, including research on genetic engineering, where the science is so complex that it may be comprehensible, if at all, only to the researchers themselves. The best that can be hoped for is that the nonscientists maintain a firm skepticism, learn how to distinguish reliable scientific informants from quacks, have the humility to admit when they are confounded, and press relentlessly until they obtain the necessary answers.

It also helps to be able to understand the scientists' perspective. I remember a talk I gave in 2006 at an invitational workshop at the American Association for the Advancement of Science (AAAS) in Washington, D.C.[45] The subject of the workshop was the ethical issues raised by human enhancement, a subject that I had been working on for a number of years. When my turn came, I described the state of the art in the use of drugs and genetic modification to give humans capabilities that were "better than normal," identified potential dangers, and outlined ways of dealing with them. I recognized most of the 25 people around the table by name, if not personally, but I had never heard of the last presenter, a thin, ruddy man with red hair and bushy eyebrows. I soon realized that he was not discussing the same types of enhancements as the rest of us, but something different: human enhancement using nanotechnology. Nanotechnology, or nanotech as it is commonly called, is a highly diverse set of technologies that share the common feature of being tiny, specifically, no larger than 100 billionths of a meter. Although still in its infancy, nanotech already has yielded some amazing products, including sunscreens containing nanoparticles of zinc oxide or titanium dioxide; medical equipment, home appliances, and cell phones that incorporate nanosilver to make them resistant to bacteria; and clothing coated with a super-water-repellant outer layer, such as the swimsuit Michael Phelps wore when he won his eight gold medals at the Beijing Olympics.[46]

I knew something about the ethics of nanotech. It didn't seem to raise any novel questions, none that hadn't already been discussed at the workshop in connection with other enhancement technologies. But the speaker did not go over any of the standard issues in enhancement ethics, such as how to protect human subjects in nanotech experiments, whether competent adults ought to be allowed to use nanotech to enhance themselves and their children, how to avoid the widening gap between rich and poor if only the wealthy could afford nanotech enhancements, or how society should value accomplishments made with the help of nanotech, such as Michael Phelps's gold medals. The speaker didn't even use the same terminology I was used to, concepts such as "autonomy," "non-malevolence," or "justice." Instead, he referred to "risk appraisal," "hazard and exposure," and "tolerance and acceptability judgment," and he talked about how organizations could develop the resilience and capacity to face unavoidable risks, the need to understand the secondary impacts of nanotech, and the extent to which a precautionary approach should be used to address uncertainty and ambiguity. I had no idea where he was coming from. Baffled, I looked down at the speaker's biosketch to see who he was. His name was Mike Roco, and he was an engineer and the senior advisor for nanotechnology at the National Science Foundation. The language he was speaking was indeed a different language than the one I was used to, the language of engineering.

The safety concerns that Roco and other nanotech engineers are worried about are broad in scope. In June 2008, a group of consumer, health, and environmental groups petitioned the Environmental Protection Agency (EPA) to stop the sale of consumer products containing nano-sized particles of silver, which is used as an antibacterial agent in such diverse products as food containers, children's toys, washing machines, cosmetics, baby bottles, and socks, on the grounds that it was a dangerous pesticide that could leech into the environment, causing harm to fish and other animals.[47] This followed a report in May of 2008 by the British Royal Commission on Environmental Pollution, which called for greater regulatory oversight over nanotech.[48] More alarming was a report by Japanese researchers who had injected mice with carbon nanotubes, a nanotech product hailed for its extraordinary strength, flexibility, and electrical conductivity. The researchers discovered that the mice developed mesothelioma, a rare type of cancer that is also produced by exposure to asbestos.[49] Then researchers in Edinburgh and Washington, D.C., claimed that carbon nanotubes produced inflammation and granulomas when injected into the abdominal cavity of mice, again similar to what happens when mice are injected with asbestos. (Granulomas are bundles of white blood cells that surround and wall off foreign objects in the body that the white blood cells cannot destroy.) In 2009, the Friends of the Earth warned that nanosilver might be a danger not only to fish and other animals but also to humans, because it was probably poisonous and could lead to the proliferation of antibiotic-resistant bacteria.[50] That same month, the EPA announced that manufacturers of carbon nanotubes would have to alert the agency 90 days before introducing a new use for the material.[51] (This requirement was later put on hold when a lawyer for the nanotech industry objected to the special procedure that the EPA had used to adopt it.)

In fact, when nanotech was in its infancy, it triggered fears that it could cause a worldwide calamity, similar to the atomic bomb, particle colliders, and recombinant DNA. These anxieties were kindled in a 1986 book called *Engines of Creation*, written by MIT-trained engineer Eric Drexler, who is often credited with having invented the word "nanotechnology" (although apparently it had been used earlier by a Japanese science professor). In his depiction of the future of nanotech, Drexler described a new manufacturing technique called molecular manufacturing, in which tiny machines would be created with the ability to reproduce themselves and to build more complex machines from the atomic or molecular "bottom" up. (This possibility was first suggested by Nobel Prize winner Richard Feynman in a 1959 lecture "There's Plenty of Room at the Bottom.") Drexler predicted that enormous benefits could result from these machines, such as tiny robots that could travel through the body and do everything from destroying cancer cells to repairing DNA. But he also exposed a corresponding threat: if they reproduced themselves without check, these minute machines could invade and take over all the ecological niches on the planet: "'Plants' with 'leaves' no more efficient than today's solar cells could out-compete real plants, crowding the biosphere with an inedible foliage. Tough omnivorous bacteria could out-compete real bacteria: They could spread like blowing pollen, replicate swiftly, and reduce the biosphere to dust in a matter of days. Dangerous replicators could easily be too tough, small, and rapidly spreading to stop."[52] In the end, the machines would consume all of the planet's resources, leaving behind what Drexler called "gray goo."

Drexler's "gray goo" scenario attracted a lot of attention. A 2000 article in the *International Herald Tribune* proposed creating "blue goo"—"tiny self-replicating police robots that keep the other ones from misbehaving."[53] Michael Crichton depicted gray goo as the "monster" in his 2002 science fiction novel *Prey*.[54] Prince Charles was said to have been so upset by the goo risk that in 2003 he convened a "nanotech summit" at his country estate in Glouchestershire,[55] although the prince later denied the report, saying that beliefs that "self-replicating robots, smaller than viruses, will one day multiply uncontrollably and devour our planet . . . should be left where they belong, in the realms of science fiction,"[56] which no doubt was a reference to Crichton's recently published novel. Meanwhile, nanotech researchers, worried that fears of gray goo would stopper their funding, rounded on Drexler and claimed that molecular manufacturing was a scientific impossibility. One of Drexler's chief critics was Nobel Prize–winning chemist Richard E. Smalley. According to Smalley, molecular manufacturing could not take place at the atomic level because of two problems: "fat fingers" and "sticky fingers." The electronic bonds that hold atoms together are sensitive to all the other atoms in the area, Smalley argued, so in order to manipulate the atoms they were working with, molecular machines would need additional manipulator "fingers" for each of the atoms. But there wouldn't be room for all these "fingers" at the atomic level; they would be too "fat" to fit. Moreover, the atoms that were being moved around would stick to the atoms of the manipulator fingers, and there would be no way to "unstick" these sticky-fingered atoms so that they could be placed where they were needed.[57]

Are Smalley and the other scientists correct in saying that we have nothing to fear from gray goo, just as scientists before them dismissed concerns about the atom bomb, particle accelerators, and recombinant DNA? Or does nanotech need stricter regulation to prevent a gray goo catastrophe? Once more, we encounter the problem of where to get knowledgeable, objective information about complex scientific issues. On the one hand, there is reason to be skeptical about the reliability of information both from nanotech scientists, who have a vested interest in seeing nanotech go forward, and from non-nanotech scientists, who may feel threatened by the success of nanotech. On the other hand, the complexity of nanotech science underscores how unlikely it is that lawyers with a "basic" college-level science competence can keep us safe, as Richard Posner suggested.

Cass Sunstein, incidentally, takes a different tack from Posner in considering potentially calamitous scientific undertakings. Sunstein takes aim at the "precautionary principle," which counsels restraint in the face of scientific uncertainty, and which places the burden of proving that an undertaking is safe on its proponents. Sunstein argues that the margin of safety that the principle requires before a risky project can be allowed to proceed is too wide and would unnecessarily deprive society of important benefits. In its place, Sunstein recommends an approach that more evenly weighs both the benefits and the risks of a scientific undertaking and places the burden on critics to show that an undertaking is unsafe in order for it to be stopped. But Sunstein does not give any suggestions for how to obtain the necessary scientific expertise to be able accurately to assess risks and benefits, and his appointment by President Obama as the director of the White House Office of Information and Regulatory Affairs was criticized by University of Texas law professor Thomas McGarity, who warned that "Sunstein will take quantitative risk assessments from people who have a clear ax to grind and treat them as gospel."[58]

Concerns about the risks from nanotech raise questions about not only how much we can rely on scientific experts to provide reliable safety information but also what approach we should take in evaluating risks created by scientific enterprises such as evolutionary engineering. Since we're talking about "engineering" evolution, should we employ Roco's nanotech engineering perspective, which sounded so foreign to me when I heard him speak at the enhancement meeting in Washington, D.C.? What would this perspective look like if it were applied to evolutionary engineering? Would it conceive of evolutionary engineering essentially as a "big science" project, a single, huge undertaking like building a dam, and downplay the extent to which it would be a collection of small steps taken by a large number of people? This difference is important because it is much harder to control lots of different actors than a smaller number of scientists involved in a single enterprise, especially one that is dependent on government funding, like much U.S. scientific research. Viewing evolutionary engineering as a single project would risk creating the illusion that it can be easily controlled or, if necessary, stopped altogether, such as when the Obama administration cut off funding for the F-22 Raptor stealth fighter.

Roco's nanotech perspective might also place a lower value on the lives that could be harmed by genetic engineering, since these lives might be thought of primarily in the abstract, that is, as "statistical" lives. Statistical lives are faceless and impersonal, like "the number of additional highway deaths that would result from an increase in the speed limit," or "the people who would be killed if a bridge were not designed to be strong enough." Roco sometimes sounds as if he favors such an approach. In describing the risks of nanotech, for example, he emphasizes its effect on "human biosystems,"[59] the components of which are not "people," but "items" and "objects." (It is also noteworthy that Posner's efforts to calculate how much should be spent on avoiding catastrophes also take a "statistical" approach when placing a value on the deaths that would result.)

A different way to value life is in terms of "identifiable" lives. This method focuses attention on the harm to actual persons, rather on harm to persons in the abstract. An illustration of the difference in perspective can be seen in how the National Highway Traffic Safety Administration (NHTSA) and a court of law deal differently with highway accidents. One cause of accidents is big trucks that cannot stop quickly enough. The NHTSA analyzed accident reports and determined that 655 people a year die in the resulting collisions, so in July 2009, it adopted a new rule requiring tractor-trailers traveling at 60 miles per hour to come to a complete stop in 250 feet, as opposed to meeting the old standard of 355 feet. According to the agency, this would save 227 lives a year and prevent 300 serious injuries.[60] These lives are "statistical" lives; in issuing its rule, the agency had no idea who the 227 people were whose lives would be spared every year, nor did it particularly care who the 655 people were whose lives every year had been lost. If a court were handling a lawsuit filed against a trucking company by the family of one of the accident victims that the NHTSA had in mind, on the other hand, members of the family likely would be in the courtroom, and some of them would probably be called to testify about the victim's life and how the accident affected them. The family's lawyer might even show photos or video clips of the victim, since the lawyer's job is to make the victim and the family "identifiable" persons, rather than abstract, faceless individuals. Why does the lawyer do this? Because lawyers want to recover as much money as possible in damages, and they assume that judges and juries will place a higher value on the loss of an identifiable as compared with a statistical life.

Another way to appreciate the different ways in which we value human life is to think, on the one hand, about how much money you would be willing to donate to an organization seeking to reduce boating accidents in which passengers drown after being tossed overboard and consider, on the other hand, the lengths to which most people would go to save drowning victims, even victims who were total strangers. The one time I encountered a drowning victim, I ended up jumping into the Potomac River and diving repeatedly down 15 feet to the muddy bottom until I located him by feel. When I admitted to a TV newscaster afterward that I wasn't a particularly strong swimmer and had never received lifeguard training, she asked me why I did it. Hadn't I realized that I could have been swept away by the river's notorious undercurrents, or knocked unconscious by a panicky victim? Being afraid never occurred to me, I replied. A life had been at stake, and I hadn't given a thought to the potential cost to myself. I didn't use the term "identifiable" to describe the victim's life, but that was how I had valued it.

Engineers like Roco of course also worry about risks to real people. When they design a bridge or electrical system, they understand that their work affects the people who will drive over the bridge or receive electricity from the system, rather than being just about conductive properties or concrete. Nor would it make sense to be concerned only with the harm that evolutionary engineering might cause known individuals and ignore the risks to unknown people who will be born in the future, not to mention to humanity as a whole, as "statistical" an entity as there can be. At the same time, we must be careful to pay close attention to the human lives that will be most directly affected. This is clear when we understand who would be most directly affected by faulty evolutionary engineering—who would be born, for example, with serious physical or mental defects. They would literally be our children. In assessing the risks posed by directed evolution, then, a good place to start is with the potential harm that these children might suffer.

PART TWO

The Hazards of Evolutionary Engineering

CHAPTER THREE

Physical Harm to Children

How dangerous would it be to try to manipulate the genes of children directly in an effort to control human evolution? We can get some idea of the risks involved by looking at what transpired when we altered the genes of plants and other animals. Many of these efforts have been highly successful, producing better crops and more nutritious food. But there have been enough glitches along the way to give us serious pause.

Humans have long sought to direct the evolution of plants and animals. Around 14,000 BC, people began domesticating wild plants by collecting and replanting their seeds. The next step was to eliminate those offshoots of the plantings that displayed less desirable characteristics. Over time, this selective breeding improved the crop. As David Suzuki and Peter Knudtson point out, "without any knowledge of underlying genetic mechanisms, [our ancestors] successfully selected for crop varieties and domestic breeds by crudely channeling natural evolutionary forces toward useful ends."[1] The first recorded instance of hybridization, the deliberate crossing of genes from two plants, took place in 1720, when Thomas Fairchild introduced a flower he called a "pink," which combined the genes from a sweet william and a carnation.[2]

The impact of these early forms of evolutionary engineering on our diet has been particularly striking. According to paleontologist Peter Ward, "there are more than two hundred thousand species of angiosperms, or flowering plants, yet only ten of these provide the vast majority of human food."[3]

Animal breeding began around the same time as agriculture. Dogs were the first animals to be domesticated, followed by sheep and goats around 8,000 BC, and then cows. Horses, donkeys, water buffalos, and llamas were domesticated around 4,000 BC. (One wild ancestor of the horse, Przewalski's horse, can still be found in preserves in Poland and has never been tamed.) Fifteen hundred years later came domesticated chickens and camels.[4]

As with plants, animal domestication was accompanied by selective breeding. A recent count identified 6,379 breeds of some 30 domesticated species,[5] most of which have been refined through at least 20 generations.[6] Techniques for selective breeding now include controlled mating, examining and testing both parents and offspring to identify optimal individuals, and careful record keeping.[7] Sexologist Robert T. Francoeur describes a particularly exotic approach that was first adopted in the early 1960s, which uses "pseudopregnant rabbits as temporary incubators for the centuplet offspring of superovulated prize cows and ewes, shipping the handy bunny incubators with their precious cargo to developing nations where teams of animal breeders transferred the embryos to surrogate mothers for normal pregnancies."[8]

In the last 35 years, evolutionary engineering of both plants and animals has entered a new phase with the ability to modify genes directly through gene splicing and recombinant DNA technology. Classic plant breeding and animal husbandry modify species over time, but gene splicing and recombinant DNA make faster, more dramatic changes possible. Recombinant DNA engineering also permits genetic material from one plant or animal to be spliced into another. In this way, "transgenic" organisms can be manufactured that combine genes from different species.

Calgene introduced the first genetically modified food in 1994, the Flavr Savr tomato. It was designed to solve the problem of how to get ripe tomatoes to the supermarket without spoiling. In the warm months, grocers can obtain ripe tomatoes from local farmers and sell them quickly enough to still be fresh. But in winter, grocers in climates that do not have a year-round growing season have to sell expensive hydroponic tomatoes or ship tomatoes in from places such as California, Mexico, and Florida. If the tomatoes shipped long distances are picked ripe, the risk is that they will spoil before they are consumed. Florida growers invented a particularly unappetizing way of coping with the spoilage problem: they harvested the tomatoes when they were green and then stored the immature fruit in huge chambers until it was time to ship them. At that point, they pumped ethylene oxide gas into the storage chambers. This turned the tomatoes a pale pinkish-red so that they looked somewhat ripe, but since it did not actually ripen them, they stayed hard and did not spoil before they were sold. Unfortunately, they also did not ripen; if left long enough in an attempt to let them do so, they turn to mush. You may know these as the barely edible "golf ball tomatoes" in cellophane-wrapped plastic trays.

The Flavr Savr tomato was designed to stay ripe for long periods without spoiling. Several scientific developments made this possible.[9] The first was the discovery of the enzyme polygalacturonase (PG), which occurs naturally in tomatoes and is what produces the protein that causes them to soften when they become ripe. The second step was unraveling the process that the tomato uses to produce the PG enzyme. This requires a substance called messenger RNA, a type of nucleic acid distinct from DNA, to translate the genetic instructions for the PG protein into a form from which the protein can be constructed. Calgene researchers learned how to introduce a modified "antisense" gene into the tomato that copies the instructions backward. This prevents the enzyme that softens the tomatoes from being produced.

Despite the sophisticated science behind its development, the Flavr Savr tomato was a commercial failure. Not only did it cost too much to produce, but it encountered opposition from people worried about its safety. Much of the opposition came from Britain and other countries in Europe, where resistance to genetically modified foods like the Flavr Savr has been particularly strong.

One particular fear is that foods that are modified with genes from plants containing substances to which people are allergic might themselves become allergenic. For example, one study found that soybeans modified with DNA from Brazil nuts contained a known human allergen.[10] Although there is little evidence that anyone actually has become ill from eating genetically modified foods, people worry about what they regard as inadequate safety testing. A toxicologist with the EPA warns that "this technology is being promoted, in the face of concerns by respectable scientists and in the face of data to the contrary, by the very agencies which are supposed to be protecting human health and the environment. The bottom line in my view is that we are confronted with the most powerful technology the world has ever known, and it is being rapidly deployed with almost no thought whatsoever to its consequences." Canadian geneticist and environmental activist David Suzuki admits that, as a geneticist, he's "very excited by what's going on in terms of genetic engineering" but nevertheless bothered by the fact that "we have governments that are supposed to be looking out for our health, for the safety of our environment, and they're acting like cheerleaders for this technology, which . . . is in its infancy and we have no idea what the technology is going to do. Anyone that says, 'Oh, we know that this is perfectly safe,' I say is either unbelievably stupid or deliberately lying. The reality is we don't know. The experiments simply haven't been done and we now have become the guinea pigs."[11] Hugh Lehman, a member of the Sierra Club's Genetic Engineering Committee, agrees: "While GMOs [genetically modified organisms] are consumed widely in the United States and Canada, to our knowledge there is no systematic effort to monitor the health of consumers to detect harms from such consumption. The health of consumers may already be affected but, since nobody is investigating, it is virtually certain that such harm will go undetected for a very long time."[12]

David Schubert, head of the Cellular Neurobiology Laboratory at the Salk Institute in La Jolla, California, points to a different problem—how industry-wide effects make it difficult to preserve traditional crops once modified versions are introduced: "If some people are allowed to choose to grow, sell and consume GM foods, soon nobody will be able to choose food, or a biosphere, free of GM. It's a one way choice, like the introduction of rabbits or cane toads to Australia; once it's made, it can't be reversed."[13]

A related concern is contamination. In 2000, genetically modified foods made headlines when a small amount of genetically modified corn was discovered in corn meal that had been used to make taco shells. The corn, called StarLink by its European manufacturer, Aventis, had been modified with genes from a bacterium to protect it from the caterpillars of the European corn borer moth, which topple the corn stalks by chewing through them. The bacterial genes produce a protein called Cry9C that perforates the intestinal tract of the worms, allowing bacteria that live in their digestive tracts to escape and kill them by infection. (This gives an idea of how intricate genetic modification of crops can be: introducing a gene from a bacterium that allows different bacteria to penetrate and kill an insect that destroys a plant.) Aventis obtained approval from the EPA to use StarLink corn in animal food and ethanol manufacturing, but not in human food, since the EPA could not rule out the possibility that Cry9C would be allergenic. To protect corn for human use, the StarLink corn was supposed to be planted at a distance, yet insufficient care was taken to prevent the kernels from becoming mixed together in grain elevators. Kraft Foods ended up pulling 2.5 million boxes of taco shells off supermarket shelves, Taco Bell replaced all of the shells in its 7,000 fast-food restaurants, and Japan and South Korea banned imports of U.S. corn. Aventis had to buy back an entire crop of StarLink and defend a number of class action lawsuits brought by growers and by consumers who claimed they suffered allergic reactions. Eventually, the company withdrew the corn from the market. For its part, the EPA decided that, henceforth, it would only allow genetically modified crops to be grown that had been cleared for human food use.[14]

The StarLink episode did little to derail the genetically modified plant industry, however. By 2006, 252 million acres of transgenic crops were being planted worldwide, with 53 percent grown in the United States alone. These include herbicide- and insect-resistant soybeans, corn, cotton, canola, and alfalfa; virus-resistant sweet potatoes; and plants able to survive extreme weather.[15] The U.S. Grocery Manufacturers Association estimates that 70 percent of all the food sold in the United States now contains ingredients from genetically modified crops.[16] Proponents point to numerous benefits. Genetic modification has increased agricultural yields by over 4 billion pounds a year in the United States and saves growers more than $1 billion a year in costs.[17] Many of the savings come from the use of fewer pesticides.[18] Genetic engineering has been used to produce foods that combat disease, such as Monsanto's "golden rice," which has had four extra genes inserted to make it unusually rich in beta carotene, which helps prevent blindness and protects against infection.[19]

Plants are genetically modified for purposes other than to be better, healthier foods. Recombinant DNA can transform them into living factories for producing human drugs, such as anticoagulants,[20] and vaccines, including a vaccine against hepatitis B.[21] Plants also have been turned into environmental cleanup agents. While some plants act this way naturally, such as sunflowers and Indian mustard, which have been sown at the Chrysler complex in Detroit in order to remove lead from the soil,[22] they can be engineered to be more efficient and to target a wider variety of pollutants. In one experiment in Connecticut, for example, the addition of a gene from the intestinal tract of a laboratory monkey gave trees the ability to absorb mercury.[23] Researchers in South Africa have also engineered a tobacco plant to pinpoint the location of land mines left over from armed conflicts; the plant changes color in the presence of nitrogen dioxide, which is given off by the aging munitions.[24]

Genetic engineering involves animals as well as plants. Honeybees and silkworms are engineered to make them resistant to disease.[25] Genes are inserted into the DNA of insect pests to render them sterile, after which they are released back into the insect population, where they mate but fail to reproduce.[26] The Enviropig reduces pollution from manure runoff by digesting (and therefore biodegrading) 90-100 percent of the phosphorous in its diet, compared with only 50 percent for normal pigs.[27] Foreign genes added by recombinant DNA to crops come not only from plants but from fireflies, fish, chickens, and hamsters.[28]

What about human evolutionary engineering? Until recently, humans primarily employed mate selection and similar low-tech breeding methods as described in the introduction. Gradually, however, we have begun to use on ourselves some of the same techniques that have become common for other animals. Not long after artificial insemination was introduced for animals, the first use in humans was reported by a British physician in 1799.[29] Louise Brown, known as the first "test tube baby" because her mother's egg was fertilized in vitro, was born in 1978. By 2006, over 3 million children worldwide had been produced with IVF,[30] and IVF has now been supplemented by an alphabet soup of other assisted reproductive techniques: gamete intrafallopian transfer (GIFT), in which sperm and eggs are placed in the fallopian tubes and fertilized there, rather than in a petri dish as is the case with IVF; zygote intrafallopian transfer (ZIFT), where the eggs and sperm are fertilized in the laboratory and then some of the resulting pre-embryos are transferred to the fallopian tubes; tubal embryo transfer (TET), where actual embryos are transferred to the fallopian tubes; frozen embryo transfer (FET); ooplasmic transfer (CT), where cytoplasm from a donor egg is added to the cytoplasm of another egg to increase its chances of developing; intracytoplasmic sperm injection (ICSI), where a sperm is injected directly into the egg; intrauterine insemination (IUI), where sperm is placed in the uterus; direct intraperitoneal insemination (DIPI), where sperm is introduced into the pouch of Douglas, a space between the rectum and the uterus; fallopian tube sperm perfusion (FSP), where a lot of sperm are put into a nurturing fluid and introduced into the uterus; peritoneal oocyte sperm transfer (POST), where eggs and sperm are injected into the stomach cavity; microsurgical epididymal sperm aspiration (MESA), where sperm is obtained from the epididymis, a tube emerging from the testicles; testicular sperm aspiration (TESA), where sperm is removed directly from the testicles; testicular sperm extraction (TESE), the same thing except that the sperm is received from a chunk of testicle that is cut off; and the list goes on and on.

A number of these techniques, such as IVF, also make it possible to test embryonic cells before embryos are implanted in the womb, the technique called PGD described in the introduction. PGD was pioneered by a British doctor in 1989. In order to reduce the risk of a child being born with a genetic disorder found chiefly in males, he removed a cell from an early-stage embryo and tested it for gender so that only female embryos would be implanted in the mother's womb and allowed to come to term.[31]

UCLA professor Gregory Stock provides an illustration of the potential power of PGD to affect the characteristics of human populations. Certain genes, he notes, have been said to be responsible for most of the genetic variation in human intelligence. Suppose a number of parents created 100 embryos using IVF, tested them prior to implantation, and only implanted the embryos with the genes that were associated with greater intelligence. Stock claims that the average IQ of the resulting group of children would be 20 points higher than the general population, placing them in the top 10 percent in terms of intelligence.[32] When these children reproduced, they could repeat this strategy, which would produce children with even higher IQs, and so on.

Stock's example emphasizes that human evolution can be directed in different ways. The parents in Stock's example may decide privately to make their offspring as intelligent as possible. On the other hand, they may act because someone told them to, such as the government or some other source of authority in their lives, for example, a religious leader. As we will see, this greatly complicates any effort that might be made to manage evolutionary engineering.

In Stock's example, IVF and PGD produce dramatic changes in intelligence in a very few generations. But the amount of evolutionary change that these technologies make possible is relatively limited, since the technologies are essentially "passive," in the sense that the parents are choosing which embryos to implant based on the naturally occurring sets of genes that the embryos have inherited. Plant and animal breeders, on the other hand, have gone beyond passive genetic engineering to employ "active" genetic engineering techniques, such as recombinant DNA, to produce plants and animals with genetic mixtures that do not occur in nature.

In comparison with active genetic changes being made in plants and animals, active genetic engineering in humans is still fairly primitive. Gene splicing has been used to correct errors in genes that are then reintroduced into patients, such as in the case of Ashanti DeSilva described earlier, but many times these interventions are not successful, and even when they are, they merely produce "somatic" genetic changes, that is, changes in nonreproductive cells that are not passed on to offspring, and therefore have at most an indirect effect on human evolution.

As noted earlier, genetic therapy and other medical interventions can impact evolution by allowing people to reproduce who previously would not have survived long enough to do so. In 1960, 90 percent of infants born with spina bifida, a disorder in which the formation of the spine is incomplete, did not survive. Now 90 percent of them survive, and many have children of their own. (In one survey, 28% of teenagers with spina bifida reported having had sex, and only 16% said they used birth control.)[33] The median life expectancy for people with cystic fibrosis is now 37.[34] This permits an increasing number of women with cystic fibrosis to live long enough to have children; the first such pregnancy was reported in 1960.[35] In the past, hardly anybody born with congenital heart defects survived; now 85 percent live into adulthood.[36]

Advances like these are notable medical achievements, but their effect on human evolution pales in comparison with the potential impact from germ line engineering, that is, modifications that affect the genes of reproductive cells and therefore can be passed on to future generations. Germ line genetic engineering has never been attempted deliberately in humans, but it does take place inadvertently. A case in point is ooplasmic transfer, an assisted reproduction method mentioned earlier in which a deficiency in an egg which prevents it from successfully developing in the womb is repaired by injecting into it cellular material taken from a normal egg provided by another woman. What makes this qualify as germ line genetic engineering is that the cellular material from the other woman's egg contains genes from that woman: even though the cellular matter that is inserted comes from outside the cell nucleus, where most of the genes are located, some genes are found in structures outside the nucleus called mitochondria, and some of these mitochondria inadvertently will be injected into the defective egg. As a result, a child born from the repaired egg will inherit genes from three people instead of from the normal two: the genes in the father's sperm and in the mother's egg nucleus, as well as those from the donor's mitochondria.

The key question is, how hard would it be to produce human germ line changes intentionally? One approach would be to remove DNA from the nucleus of one of the cells in an embryo fertilized in the laboratory, use gene splicing to modify the DNA, put the DNA back into the cell nucleus, and then stimulate the cell to make it begin to divide normally. If this were done early enough in the development of human embryos, the changed DNA would show up in all of the cells of the resulting persons, including their eggs and sperm, and therefore be passed on to their children as well. Another approach that has been suggested is to construct an artificial chromosome, described earlier, embed the modified DNA in it, and inject the chromosome into an embryonic cell. In fact, as Stock points out, germ line engineering actually might be easier to accomplish than changing the genes in somatic cells, since a germ line change affects all of the cells in the body, and there is no need to make sure that the modified genes find their way to the appropriate place in the body, one of the major challenges in somatic cell engineering.[37] For example, suppose someone wanted to make a genetic modification to build stronger muscles. If the modification were not made to cells in an early-stage embryo, then the modified DNA would have to be injected into as many cells as possible in each of the muscles to be strengthened. The dismal prospect of being punctured repeatedly, perhaps thousands of times, by needles containing modified DNA is no doubt one of the reasons why athletes seeking a competitive advantage have not tried this technique, even though several genetic modifications have been identified that might do the trick, such as in the gene that codes for a protein called myostatin.[38]

Even if we learn the basic techniques for modifying human germ cells, however, this does not mean that we would be pleased with the result. Since we've been modifying animal germ cells for quite some time, you might think that we can be confident that the same techniques will work in humans. As paleontologist Peter Ward maintains, "As easily as we breed new varieties of domesticated animals, we have it in our power to bring a new human race, variety, or species into this world."[39] But it hasn't always been easy to achieve positive results in animals. Take cloning, which involves some of the same scientific methods as germ line engineering. Scientists can now clone animals from adult cells, which avoids the need to destroy embryos to obtain the starter cells, an accomplishment that might make cloning less objectionable in humans. But it took 277 attempts to clone Dolly the sheep, the first mammal to be cloned from an adult cell. Each failure resulted in the destruction of a developing embryo, and Dolly herself died prematurely, which some attribute to the adverse effects of the cloning process. Moreover, this type of cloning has not yet proven successful in primates. In 2007, researchers reported that they had cloned monkeys for the first time using adult cells, but none of the embryos survived past 25 days.[40]

Even if genetic changes "work" in the sense that the modification is made to the animal, the animal often suffers serious physical harm. Thirty years ago, broiler chickens reached their full weight in about 80 days; the old-fashioned engineering technique of careful breeding has now shaved that time in half. But the rate at which bones and the cardiovascular system grow turns out to be slower than the growth rate of the rest of the chicken, so the birds suffer leg and heart problems, and their immune systems are compromised, making them more prone to disease. What about those Thanksgiving turkeys that have such large, meaty breasts? They too have leg problems and are so heavy that the toms cannot rise up high enough to mount the hens, so the hens have to be artificially inseminated. In 1985, Congress created the National Pork Board to encourage the breeding of leaner pigs. It worked, and pork became "the other white meat," but the pigs become stressed or have heart attacks when they exercise. Cows bred to produce more milk get digestive disorders, foot rot, skin and skeletal disorders, udder edema, and teat injuries.[41]

Another genetic engineering approach is transgenics—creating animals with genes from other organisms. Many times, however, the inserted genes do not function properly.[42] One well-known incident was the attempt in the mid-1980s to insert a gene that produces human growth hormone into pigs in order to make them grow faster and less fatty. The pigs, known as the "Beltsville pigs" after the U.S. Department of Agriculture site in Maryland where the experiment took place, did in fact have less fat, but they were plagued by diarrhea, enlarged mammary tissue in males, lethargy, arthritis, lameness, skin and eye problems, loss of sex drive, and disruption of their fertility cycles. The human gene was indeed activated ("expressed") in 19 of the pigs, but 17 of these died prematurely of causes that included pneumonia, pericarditis (inflammation of the tissue around the heart), and stomach ulcers.[43] As described in an article in the *New Scientist*, "this pig [was] a thorn in the side of high-tech agriculturists and an icon for animal rights activists everywhere. . . . The engineers added a genetic switch that should have turned on the growth hormone gene only when the pig ate food laced with zinc. But the switch failed. The extra growth hormone made the pig grow faster, but it also suffered severe bone and joint problems and was bug-eyed to boot." The article proceeded to point out that, "of course, unlike human experiments, slaughtering 'failures' is always an option for animal genetic engineers."[44]

Even if an animal whose genes have been modified appears healthy, adverse effects may show up in offspring or subsequent generations. Offspring might grow too large for normal births.[45] Cancer has shown up in the descendants of genetically engineered mice.[46] And as Peter Ward points out, "all domesticated animals appear to have undergone a loss of intelligence compared with their wild ancestors."[47]

It therefore should come as no surprise that even the relatively modest attempts at genetic engineering in humans have included some noteworthy failures. The most well-known death from a gene therapy experiment was that of 18-year-old Jesse Gelsinger, who died in 1999 in an experiment that was part of a research program at the University of Pennsylvania to develop a genetic treatment for a genetic disease called ornithine transcarbamylase (OTC) deficiency, which affects one out of every 40,000 births. The livers of babies with the deficiency do not produce an enzyme that enables the body to metabolize ammonia, which is a by-product of the breakdown of protein. Affected newborns fall into a coma within 72 hours of birth and suffer severe brain damage. Within a month, half are dead; within six months later, 75 percent. Gelsinger himself did not have a full-fledged version of the disease. Instead, he had what is termed a "mosaicism," where only some of his liver cells were unable to manufacture the missing enzyme. As a consequence, Gelsinger was fortunate to survive without suffering brain damage, and he was able to control his disease by eating a nonprotein diet and taking enzyme pills. Gelsinger nevertheless was willing to participate in the genetic experiment to help find a cure for the more seriously affected infants. Moreover, Gelsinger was taking more and more enzyme pills as time went on; by the time he enrolled in the experiment, he was up to 35 pills a day, so he was interested in finding a cure for himself as well.

The Penn researchers eventually hoped to insert functional genes into the livers of seriously ill babies with OTC in the hope that it would enable them to manufacture the missing enzyme. But first they had to determine the maximum tolerated dose of the modified genes, that is, the highest dose that could be used without producing serious side effects, and also find out if the virus that they proposed to use as the vehicle or "vector" to carry the genes to the liver, adenovirus—the virus (actually, a retrovirus) that causes the common cold and one of the vectors most commonly used in gene therapy experiments—would get the modified genes where they needed to be and in good enough shape to do their job. Originally, the researchers proposed to use severely affected newborns as their subjects, but they consulted Arthur Caplan, a leading bioethicist at Penn, who persuaded them to use relatively healthy adults such as Gelsinger instead, since the adults would be better able to give voluntary informed consent to serve as subjects than the parents of dying babies.[48] There were 18 subjects in all, divided into groups that would receive different dosages. Gelsinger was in the highest dosage group.

During the night after his infusion, Gelsinger began feeling ill and running a fever, but the doctors weren't alarmed, since other subjects had experienced the same side effects. But Gelsinger's condition continued to deteriorate. He lapsed into a coma and, four days later, died of massive organ failure.[49]

Two months later, the FDA inspected the research operation at Penn, found fault with the informed consent process that the Penn researchers had used, and also determined that Gelsinger should not have been a subject because his liver was not functioning properly when he was given the experimental infusion. In 2005, the Justice Department settled a lawsuit against the Gelsinger researchers in which they were accused of making false statements to the FDA. Their institutions paid over $1 million in fines, twice the funding they had received from NIH to run the experiment. Gelsinger's father also filed a civil suit, which eventually was also settled.

Gelsinger's death illustrates not only the risks of human genetic experimentation but the ways in which economic conflicts of interest may make researchers less careful than they should be. The experiment in which Gelsinger died was part of a program at the University of Pennsylvania's Institute of Gene Therapy, at the time the largest university program in gene therapy research in the country. The director, James M. Wilson, was also the founder of and a stockholder, along with the university, in Genovo, a company that had the exclusive right to profit from any of the institute's discoveries. By the terms of the settlement with the Justice Department, Wilson was barred from conducting FDA-related research on human subjects until 2010.

Another well-known incident involving a human genetic experiment was an effort by French geneticists in 2000 to insert corrected genes into the bone marrow of babies born with a genetic condition that prevented them from developing an immune system. At first, the experiment seemed a success: the babies' immune systems began to function perfectly, and the French doctors heralded this as the first complete gene therapy cure. Shortly afterward, however, three of the children developed leukemia, a form of cancer. It turned out that the retrovirus that the researchers had used to carry the corrected genes into the children's DNA had unfortunately implanted itself too close to a cancer-causing gene, thereby activating it.

In 2007, a 36-year-old woman named Jolee Mohr participated as a subject in another gene therapy experiment. Mohr had rheumatoid arthritis, and her doctor enrolled her in a study he was running for a company named Targeted Genetics to test a novel, gene-based arthritis treatment, tgAAC94. This is a virus engineered with an extra gene that makes it manufacture proteins that counteract a substance called tumor necrosis factor-alpha, a major cause of the inflammation that rheumatoid arthritis produces in the joints of its sufferers. Mohr received two injections of the engineered virus. The first produced no adverse effects, but she became ill after the second injection five months later and died three weeks after that.[50]

A review by the NIH concluded that Mohr had died of a fungal infection that was probably not associated with the experiment, and six months after her death, the study she had been enrolled in recommenced. Nevertheless, her death, along with Gelsinger's and the leukemia in the French children, inspired *Next*, the last novel that Michael Crichton published before he died. "Do you know the history of gene therapy?" asks one of Crichton's characters, who goes on:

> It's a horror story, Ellie. Starting back in the late 1980s, the biotech guys went off half-cocked and killed people right and left. At least six hundred people we know about have been killed. And plenty more we don't know about. . . . You know how gene therapy kills people? All sorts of ways. They don't know what's going to happen. They insert genes into people, and it turns on cancer genes, and the people die of cancer. Or they have huge allergic reactions and die. These goofballs don't know what the hell they are doing. They're reckless and they don't follow the rules.[51]

Since Crichton's book is supposed to take place in the present, his character's account is grossly exaggerated. The 600 deaths to which the character refers are actually 691 "serious side effects" from gene therapy experiments that came to light in 1999 following an investigation by the

NIH.[52] Only six of these cases resulted in death,[53] and we know of less than a dozen confirmed cases all told in which people have died in gene experiments. There may be others that are being kept secret—of the 691 serious side effects discovered by the NIH in 1999, only 39 had been reported to the NIH as required by law[54]—but the number of deaths clearly is nowhere near the 600 in Crichton's novel.

Still, Crichton's book touched a nerve, and failed experiments have continued to raise doubts about whether gene therapy is safe enough to be tested in humans. One concern is unanticipated harms. While Gelsinger's death evidently was not that surprising given the investigators' failure to conduct adequate tests to see if his constitution was robust enough to withstand the cold virus that they injected into his system, the leukemia in the French babies was completely unexpected. Would children whose genes were modified by their parents suffer unforeseen harms, perhaps harms that did not show up until the children grew to adulthood? One reason for concern is the limited ability of animal testing to predict long-term health effects in humans, for example, whether a patient cured of a fatal nerve disease might develop a severe nerve disorder 20 or 30 years later.[55] Germ line engineering is especially worrisome. As bioethicists Emily Marden and Dorothy Nelkin point out, "the ultimate safety of germline genetic manipulations may be unknowable until many years after the treatment."[56] Some adverse effects might not become evident until engineered children had children of their own, or even later, after many generations. Even fairly routine genetic technologies might be problematic. Some studies have found that IVF increases the risk of premature birth and lower birth weights.[57] This is significant not only because of the many children that have been conceived using IVF but because the technique is probably necessary in order to make germ line changes, which have to take place at an early stage of embryonic development in order to be passed on to future generations.

Why is human genetic engineering so fraught with difficulty? One reason is that genetics in general and human genetics in particular are full of surprises. Until a decade ago, scientists believed that there were approximately 100,000 genes in human DNA. When the human genome was finally sequenced, the number turned out to be closer to 25,000. How could so few genes account for such a complicated organism as a human being? For that matter, how could a tiny member of the cabbage family called thale cress, the first plant to have had its DNA decoded, turn out to possess roughly the same number of genes as humans? The explanation, discovered only recently, is that the same gene can perform many different functions, depending on where and when it is turned on and off. The controls for activating and deactivating the genes are a set of genetic switches called "transcription factors" and "enhancers." These switches are located in the long stretches of the DNA molecule that lie between the genes. Until a few years ago, these regions of DNA were called "junk DNA" because scientists were certain that they did not play any functional role. Now, geneticists who are conducting "genome-wide association studies," in which they scan the DNA of many individuals in order to identify common variations associated with various traits, are finding 80 percent of these genetic variations in the "junk" DNA between the genes;[58] these regions of DNA have now been relabeled "regulatory."

Like most aspects of genetics, the switching system is extraordinarily complex. Not only can genes be switched on and off, but the functions of a single gene can be affected by many different switches. To gain an idea of just how complex the switching system is, consider this: In order for a gene to work, that is, to produce a protein, one of the things that has to happen is that an enzyme called polymerase has to trigger something called a core promoter, located in the regulatory region next to the gene. The production of polymerase is controlled by a series of "enhancers" and "silencers." These are sometimes located in far distant regions of the DNA molecule, and each of them can affect a number of different genes. The enhancers and silencers influence the production of polymerase through the actions of large groups of proteins called "activators" and "repressors," which are relayed through another set of proteins called "coactivators" and "basal factors." Once the polymerase triggers a core promoter, the core promoter converts or "transcribes" the DNA in the gene into a substance called messenger RNA. As many as fifty different proteins may be involved in transcribing the DNA in a gene into messenger RNA, with each protein produced by different genes, and the production of each protein triggered by a similar complex set of processes.[59] Still with me? Now consider that this is only one of many steps that must take place in order for a gene to do its work.

The ability of the same gene to perform many functions as a result of being turned on or off is called pleiotropy, and it is one of the main reasons why it may be perilous to modify human genes. A mistake would impact not only the particular trait being targeted, such as a genetic disorder, but all the other traits that are associated with that gene. Mistakes are likely to be particularly dangerous when they occur in the course of adding DNA to supplement the functioning of a normal gene by making more of an existing protein, as opposed to being introduced to provide a protein that is missing because of a defective gene. As explained by W. French Anderson, the geneticist who performed the first successful gene therapy experiment on Ashanti DeSilva, adding protein "might adversely affect numerous other biochemical pathways. In other words, replacing a faulty part is different from trying to add something to a normally-functioning, technically complex system."[60]

Even merely trying to correct genetic malfunctions may produce untoward results if not enough is understood about how genes work. The classic example is the genetic error that causes sickle cell disease. The error is a single incorrect nucleotide, the basic building block of DNA, amid the 3 billion that make up the entire human genetic code. If geneticists were able to cut out the disease-causing nucleotide in viable embryonic cells and substitute the correct one, they might be able to prevent the children who developed from those embryos from being afflicted with sickle cell disease. But 50 years ago researchers discovered that in certain parts of Africa the sickle cell trait provides resistance to malaria, and the children whose DNA had been corrected would lose this resistance. Undoubtedly other "bad" genes confer as-yet-unknown benefits that might be forsaken if the genes were removed from human DNA. One commentator wonders, for example, what would happen if any "intellectually desirable attributes are also transmitted with the complex of genes responsible for schizophrenia."[61]

Scientists who nevertheless are keen to try their hand at genetic engineering argue that the discovery of genetic switches provides the solution to the problems caused by pleiotropy. To avoid upsetting all of the functions associated with a gene, these researchers propose to only modify the switches that affect the targeted trait, not the gene itself.[62] But given the intricacies of the switching mechanism, it is not clear how easy this would be to accomplish, and bioethicist Nicholas Agar thinks that trying to reengineer the switching system could end up restoring much of the randomness in natural evolution that genetic engineering is designed to counteract. "Changes to imperfectly understood complex systems," he observes, "produce effects that, relative to our knowledge, are random." Moreover, random changes are not neutral in terms of their effect on an organism: "Random changes to complex, well-functioning systems," notes Agar, "are much more likely to make them work worse than better."[63]

Geneticists also recently have begun to understand that environmental factors, such as diet, exposure to damaging substances, stress, activity, and nurturing, not only play a large role in health and behavior but can affect whether and to what extent genes are turned on or off in specific tissues. This understanding is part of the emerging field of "epigenetics," the study of changes in heritable traits that are produced in an organism by mechanisms other than changes in their DNA. (One field of evolutionary biology that calls itself "evo-devo" emphasizes the importance of epigenetic effects that take place during the early stages of development, such as in the uterus or early childhood.) In other words, environmental factors can modify the functioning of genetic switches, and as a result, tinkering with the switches themselves may not produce the intended effect because of conditions in the environment. In addition, there is evidence that epigenetic effects can be inherited.[64] Therefore, any unintended results introduced by the operation of environmental factors may be passed on to future generations.

Finally, genetic modifications that seemed like a good idea at the time may turn out to be harmful because of unexpected changes in the environment. Some parents may prefer to have children with fairer skin for social or aesthetic reasons, for example, but if global warming unexpectedly accelerated, light-skinned children might be at a disadvantage compared with darker-skinned children, who would be less likely to develop skin cancer from excessive solar radiation.

Those at immediate risk of harm from genetic engineering are not only the children whose genes are sought to be engineered and their descendants, but their parents. Some techniques for altering germ cells involve tinkering with eggs or sperm while still in the ovaries or testes. As Susannah Baruch and her colleagues at the Genetics and Public Policy Center at Johns Hopkins point out, this could damage the cells that yield the mature reproductive cells and cause infertility, which already has taken place when these approaches have been tried on animals.[65] Mothers also might be placed at risk during labor by modifications that physically altered the child. For example, greater intelligence in animals is associated with larger brains, so it is possible that increasing intelligence in humans would be accompanied by an expansion of our brains. Nick Bostrom and Anders Sandberg actually seem to advocate this, envisioning "interventions that moderately increased brain growth during gestation."[66] But bigger brains require bigger heads, and this could prove hazardous during the birthing process as infants with enlarged heads tried to make their way down the birth canal.

The problem stems from the fact that the way humans deliver babies is a highly imperfect evolutionary compromise, known as the "obstetric dilemma."[67] On the one hand, our brains, and hence our heads, are larger than those of other primates. On the other hand, the birth canal in humans is shorter and narrower than in other primates in order to strengthen the pelvis so that it can support our upright posture. In order to get their big brains (and shoulders) through the narrow pelvic passage, human infants must rotate themselves, and so they exit facedown, in contrast to other primate babies, which are born faceup. As a consequence, while other primate mothers can grasp the newborn as it exits and lift it to suckle, it is almost impossible for human mothers to give birth without the assistance of someone else to catch and reposition the baby. In addition, the contest between head size and pelvic shape makes the birthing process much more painful and dangerous for humans than for other primates. One researcher at the State University of New York at Stony Brook estimates, for example, that between 20 and 25 percent of all human births have ended in the death of the mother or the fetus.[68] The toll would be much greater except that human babies are born "prematurely" in comparison with apes and monkeys—17 months earlier in terms of development than chimpanzees, for instance;[69] if not for this prematurity, babies' heads would be even larger, and the risks to mother and baby even greater.

One response to the obstetric dilemma is to deliver babies surgically via cesarean section, which now accounts for 30 percent of all U.S. births. Surgical delivery has risks of its own, however, including infections, unplanned hysterectomies, and embolisms.[70] These risks are bound to be reduced with medical progress, and genetic engineers might even be able to refashion human pelvic development so that natural births became easier, but the point remains that genetically modifying one trait may require a cascade of other changes, each with its own set of dangers and unforeseen consequences.

In short, attempts to genetically engineer humans create the risk of serious and even deadly physical effects to the children who are manipulated and their descendants, as well as to their mothers. But children can suffer other types of damage besides physical injury. What kinds of nonphysical harm might be caused by efforts to direct human evolution?

CHAPTER FOUR

Psychosocial Harm to Children

There is always the chance that faulty genetic engineering could inflict physical injuries on children. But even if the engineering worked, the desired modifications were successfully installed, and there was no immediate or obvious physical harm to the child, the child still might be worse off than if he or she had not been intentionally redesigned. In fact, some opponents of evolutionary engineering oppose changes that are likely to produce benign or beneficial effects on the basis that the lives of future persons would have been engineered without their consent.[1]

Consent, or more specifically "informed consent," is indeed a fundamental principle of biomedical ethics, and future persons obviously cannot give theirs in advance. Moreover, as the Council for Responsible Genetics points out, future persons "harmed or stigmatized by wrongful or unsuccessful germ line modifications" would likely have no recourse against the ancestors who were responsible.[2] Yet few commentators are troubled by the absence of consent or accountability. Bioethicist Ray Moseley argues that "taken at face value this argument would imply that it is unethical to do anything affecting future generations, including produce them, since one cannot acquire their permission nor predict their wishes." Moseley believes that the consent problem can be overcome if we simply avoid germ line changes that are predicted to lead to bad consequences.[3] The late Marc Lappé, although a vocal crusader against genetically modified foods, points out that "future generations" includes the next one, and that "to reject all germ line alterations as 'unethical' because not all germ line–engineered individuals will be assured normal protective options or that they may be genetically unsuited to future environments is tantamount to saying no one should have children."[4] Theologian Ian Barbour suggests that "widespread public approval" can replace lack of informed consent by future generations.[5] Nicholas Agar proposes that genetic enhancement be permitted unless it made it unlikely that the child could lead a successful life "founded on values that oppose those of the enhancers."[6] So parents presumably could install artistic talents so long as doing so did not prevent their descendants from successfully pursuing nonartistic careers, say, in investment banking.

Yet how likely is it that parents will make correct choices about which traits to modify? Modifications that parents think would be advantageous might turn out to be social or economic handicaps. Parents might think it desirable to install artistic talents in their children, for example, but when the children reached adulthood, they might be unable to make a living as an artist if the economy suffered a severe, long-lasting economic recession, or even just because too many parents had made the same choice and the art market had collapsed under the weight of too many objets d'art for sale.

Aside from the difficulty of predicting which traits will be socially advantageous in the future, genetic choices might turn out to be unwise in that difficulties created by genetic changes could end up greatly outweighing any advantages. British biologist Michael J. Reiss is one person who is concerned that genetic changes could end up causing children more harm than good. He cites the example of a genetically engineered strain of fruit flies that learns ten times faster than normal flies. "At first sight," comments Reiss, "the application of this technology to humans sounds marvelous. Imagine learning ten times faster; think of all the benefits it could bring." But, he adds, "improved learning implies improved memory and if you have a far superior memory you will forget far less. Most of us have experienced unpleasant happenings which we are only too grateful to forget."[7]

Consider one of the chief goals of the transhumanists, to substantially extend the human life span. Longevity could create serious social problems, such as overpopulation and stress on the environment. Careers might last longer, but there might be fewer entry-level openings and less upward job mobility. Longer-lived persons might consume more societal resources, putting financial pressure on government programs such as Medicare and on communal services such as public safety and transportation, and competition for scarce resources could increase the conflict between young and old. Life extension also could destabilize family relationships. Longer-lived individuals could grow bored with longer marriages, increasing the divorce rate. Successive marriages could weaken the emotional bond between parents and children. If older persons continued to be able to reproduce, siblings could become more estranged as the gaps in their ages widened. Inheritances would be more meager if parents consumed more of their wealth over their longer lifetimes. In short, people might live longer but be less happy.

Other genetic changes could produce similar unwelcome trade-offs. Some of the Sleepless children in Nancy Kress's novel *Beggars in Spain* drive their parents crazy by not sleeping, and one of the original 20 Sleepless babies is shaken to death by its sleep-deprived mother; even if parental misconduct stopped short of physical violence, children who were especially difficult to raise might suffer from lack of love. At the same time, children who were made extremely beautiful could suffer greater mental anguish than more ordinary-looking people when their appearance deteriorated with age. While it seems self-evident that, if feasible, parents should engineer their kids to be smarter, could the children become, as the saying goes, too smart for their own good? The psychological difficulties currently encountered by child prodigies, for example, are well documented. These children can be acutely afraid of failure and more adversely affected by developmental disorders such as dyslexia. They may do poorly in school because they are bored or because their extraordinary abilities in some areas, such as verbal intelligence, are accompanied by deficits in others, such as spatial reasoning; this imbalance can even make their average performance so mediocre that their talents go unrecognized. Prodigies also are known to develop a "false self," a radical distortion of their personalities, in order to avoid provoking the slightest signs of parental disapproval, something to which they are often extremely sensitive.[8] And when they grow up, they often have adjustment problems when they see the abilities of their peers beginning to catch up with their own.[9] Conceivably, genetic engineering in the future will be able to prevent developmental disabilities and correct any undesirable mental states in the supersmart, but the development of these engineering capabilities may lag behind the ability to engineer greater intelligence, leaving the smart children vulnerable to the afflictions currently suffered by child prodigies.

Decisions about what genetic modifications to make also are likely to be culturally driven, which means that they will be culturally biased. In a report for the AAAS, for example, Mark Frankel and Audrey Chapman worry that social pressure to modify children "may promote something analogous to a kind of 'soft eugenics,' a 'kinder, gentler program to "perfect" human individuals by "correcting" their genomes' in conformity to specific societal norms or to an identified 'economically successful genotype.'"[10] Oxford philosopher Jonathan Glover quotes a comment by an Australian colleague that if it had been possible to engineer character traits in the Victorian era, the aim would have been to make children "more pious and patriotic."[11] Gregory Stock points out that "if you wanted to build a superior human, you would probably choose black skin, at least if the person was going to spend much time in the sun. . . . But don't hold your breath waiting for such engineering. A genetic module for black skin is unlikely to be in big demand anytime soon. Blacks certainly don't need it, and few nonblacks will want it."[12] Many Asians have their eyes surgically altered by placing a crease in the upper lid to give them a Western appearance; many of them might choose to make this change in their offspring through germ line genetic engineering if it were possible to do so. Kids likely would be made taller, since greater height commands a societal premium in most parts of the world. In the United States, for example, sperm banks turn away short men who wish to be donors,[13] and some parents whose children are of normal height are reported to seek injections of human growth hormone so that the kids can grow up to be star basketball players.[14] In China, being tall is so advantageous that the government has resorted to banning a type of cosmetic surgery in which a person's shin bones are repeatedly sawn apart and fixed in place with metal pins and braces, which adds an inch or two each time when the ends of the bone fuse together across the gap.

Perhaps there is no great harm from using genetic engineering to change the shape of people's eyes or to make them taller, but what if the Chinese still practiced foot binding, the historical practice in which girls' feet were grossly deformed to make them more sexually alluring, and parents wanted to engineer their daughters' feet this way? Or what if tastes changed? Parents in India overwhelmingly prefer male children, and they already use biomedical technology to control the sex of their children. Thus, there are more than 250 clinics in Bombay that test fetuses for gender, technicians with portable ultrasound machines go around offering their services in rural areas, and a study of 8,000 Indian abortions showed that in all but one the aborted fetus was female.[15] But what if for some reason cultural preferences changed in the course of an engineered person's lifetime and being male, for example, became stigmatized instead (not to mention the risk, discussed later, that large gender imbalances could imperil the survival of the human genetic line)?

Jonathan Glover offers his own answer to these "what-ifs" by declaring that there is no reason to assume that "opting for the genetic status quo involves less commitment to a world view than opting for a change."[16] Opponents of genetic engineering often are conservative and theistic, for example. Thus, Bill McKibben, the author of *Enough*, maintains that using genetics to improve the human race would violate our religious heritage. "In the Western tradition," McKibben explains, "the idea of limits goes right back to the start, to a God who made heaven and earth, beast and man, and then decided that it was enough, and *stopped*" (emphasis in original).[17] When Harvard professor Michael Sandel opposes genetic enhancements on the grounds that they negate "the gifted nature of human powers and achievements,"[18] it doesn't take much to figure out Who he thinks is the gift giver. Leon Kass, the former chair of President Bush's Council on Bioethics, makes a similar objection to human cloning when he argues that it is "a major alteration, indeed, a major violation, of our *given* nature" (emphasis added)[19] and when he scoffs in reference to genetic engineering that "man, even at his most powerful, is capable only of playing at being God."[20] Yet the fact that McKibben, Sandel, and Kass share a desire to preserve certain religious values does not negate a concern for children who were engineered in ways that were once fashionable when fashions change. Given prevailing notions of female beauty, for example, parents might opt to produce girls who were thin and boyish. A backlash against this body type, however, might result in these women being shunned as disturbing reminders of a misguided era of self-loathing and eating disorders.

Decisions about what traits to engineer may be not only culturally driven but dictated by the views of dominant economic and political elites. After all, it is these families that will be most able to afford the high costs of genetic engineering. Even the rather primitive technologies so far available, such as selective implantation of embryos produced through IVF, are extremely expensive.[21] For example, it currently takes multiple cycles of IVF to produce a single viable pregnancy, at an average cost of over $12,000 per cycle,[22] and this does not include the cost of the pre-implantation genetic testing needed to identify the desirable embryos to be implanted. Would-be parents thus have to spend in the neighborhood of $50,000 to take advantage of the relatively unsophisticated genetic engineering options now available, which is equivalent to the median household income in the United States.[23] The price of genetic engineering is bound to come down over time as the technology is refined, but for the foreseeable future, it will be unaffordable except for the well-to-do. In a later chapter we will discuss the implications for liberal democracy. But even if prices dropped, elites are likely to continue to shape public attitudes toward genetic engineering through their role as opinion leaders, and this could adversely affect children as parents sought to engineer them in ways that would increase their chances of joining the ranks of the elite. If genetic factors associated with narcissism were ever discovered, for instance, the success of people like Donald Trump and Oprah Winfrey might lead some parents to engineer their children to be more narcissistic. But this might not make the children socially successful. As one Manhattan psychiatrist points out, "it sounds more impressive to say that someone is a narcissist rather than a jerk."[24]

Do you recall the observation in chapter 2 that it will be difficult to regulate evolutionary engineering because to a large extent it will comprise many individual actions? Law professor and bioethicist George Annas thinks that fragmented decision making of this sort is another reason why people might make poor decisions about what traits to genetically engineer. Instead, he urges us to use the political process to decide as a matter of public policy how to go about evolutionary engineering. "Will we as a society permit individual scientists to try any or all of these experiments on humans," he asks, "or can we learn from the unanticipated consequences of conquest and the horrors of war, that humans are better off when they think before they act, and act democratically when action can have a profound impact on every member of the species?"[25] Not long after he wrote this, Annas proposed an international treaty to ban human germ line genetic engineering, removing these types of decisions from parents and clinicians and placing them in the hands of government. Later we will discuss whether Annas is correct in believing that collective decision making can best avoid misguided decisions by private individuals. Annas's treaty was never adopted, however, raising questions about the feasibility of his approach.

Even if genetic engineering did not injure a child physically or unintentionally give it an unfavorable set of genetic attributes, the child's quality of life could be compromised simply because of how others reacted to the fact that it had been engineered. For example, the idea of children being "designed" rather than brought into the world the "natural" way has led to worries about the potential adverse impact on the child's relationship with its parents. Jonathan Glover thus states that "too much genetic intervention might make us feel ourselves mere puppets of our parents and the technology they had at their disposal."[26] Critics of genetic engineering speculate that the children would be "commodified," that is, regarded by their parents less like persons and more like "things." They point to the flourishing industries that already have sprung up to serve parental demand for designer children, from adoption services that employ genetic testing to donor eggs and sperm, surrogate wombs, and fertility clinics, and urge us to imagine how much more "manufactured" children might seem as increasingly advanced forms of genetic engineering enter the market. Leon Kass, for example, worries that "increasing control over [a child's genetic makeup] can only be purchased by the increasing depersonalization of the entire process and its coincident transformation into manufacture. Such an arrangement will be profoundly dehumanizing, no matter how genetically good or healthy the resultant children. . . . The commodification of nascent human life will be unstoppable."[27]

One result of commodification could be that parents would be less accepting of their children's shortcomings, viewing them as design flaws rather than as the luck of the genetic draw. President Bush's Council on Bioethics thus warns of a shift in parental perspective "from seeing a child as an unconditionally welcome gift to seeing him as a conditionally acceptable product,"[28] while the AAAS envisions parents evaluating their children according to "standards of quality control."[29]

An example of what might seem like a product-like mentality toward engineered children is a lawsuit by parents against the fertility clinic at the University of Utah.[30] When David Harnicher's sperm proved unable to fertilize his wife Stephanie's egg, even with the aid of artificial insemination, the couple sought help from fertility specialists at the clinic, who came up with a three-pronged approach. First, they would harvest eggs from Stephanie and drill holes in them, in effect creating artificial doorways to make it easier for sperm to enter the eggs. Next, they would sell to the Harnichers sperm from a donor who matched David's appearance. Finally, they would mix the donor sperm with David's and put this in a petri dish containing Stephanie's eggs. The goal was not only to enable the Harnichers to have a baby but to enable them to represent the baby as being biologically their own, since it would look like David, and unless a genetic test were conducted to ascertain whose sperm had actually fertilized the egg, there would be no easy way to tell that it wasn't his.

The IVF worked, the Utah fertility doctors implanted the fertilized eggs in Stephanie's womb, and as often occurs with IVF, Stephanie gave birth to triplets. At first, she and her husband were delighted, but then they noticed that unlike David, who had dark hair, one of the babies had red hair. It turned out that the fertility clinic had mistakenly sold the Harnichers sperm from the wrong donor. The Harnichers thereupon sued the clinic, claiming that they had sustained emotional injury from their "disappointment" in their children's appearance and from the fact that children who had been conceived with sperm from the correct donor would have been "better looking." In a split decision, the Utah Supreme Court rejected the parents' claims, ruling that the parents could not prove that they in fact had been harmed. But the court didn't seem to be troubled by the parents' contention that the children's quality was not up to their expectations. Indeed, the judges noted that the parents had not asserted "any racial or ethnic mismatch between the triplets and their parents," suggesting that if the Harnicher babies' appearance had been different enough from what the parents had expected, their case might have succeeded.

Another concern with engineering one's children is that parents who tried to have a direct hand in their children's genetic makeup might feel personally responsible if their children did not turn out as planned. Parents often feel responsible for their children's faults, but the adverse psychological effects could be more extreme if parents felt less able to shift as much of the blame to the roll of the genetic dice. The resulting guilt could make parents depressed and embittered, further souring the parent-child relationship, and interparental recriminations could lead to marital discord and dissolution. Not only would this threaten children with less stable family environments, but the children might well believe that they were the root cause of the family's problems and so suffer the exquisite psychological torment that this attitude can inflict.

Even if parents were pleased with the "product" that emerged at the end of a baby assembly process, the high cost of the genetic modification process might so impoverish the family that the children were deprived of other important resources, ranging from family vacations to healthy food. Parents also might not be able to afford to modify all of their children or to modify all of them equally, resulting in corrosive levels of sibling rivalry.

Parents who designed their children also might harm them by confining them to narrow developmental pathways. Parents who altered their children's genes to make them exceptionally smart, for example, might insist that they become intellectuals rather than athletes or artists. Children given martial talents might be expected to continue the family's tradition of military service. Parental pressure of this sort would seem to deny children what philosopher Joel Feinberg called the "right to an open future. "It is a duty of parents," Feinberg proclaims, "to keep as many as possible of a child's central life-options open until the child becomes an autonomous adult himself, and can decide on his own how to exercise them."[31] Children whose parents confined them to predetermined paths may lead less happy lives because they feel overcontrolled, stunted, and unfulfilled.

Not everyone agrees with Feinberg, however. Asserting what he calls the "Principle of Procreative Beneficence," Oxford professor Julian Savulescu states that, rather than allowing their children the freedom to chart their own directions, parents have a duty to produce a child who "is expected to have the best life, or at least as good a life as the others, based on the relevant available information."[32] While conceding that self-creation and independence are worth fostering, Jonathan Glover argues that they are impossible to achieve: "The influence of our genes on the choices we make excludes total self-creation. The influence of the parenting we are given excludes both total self-creation and total independence."[33] Others reject the notion that parents would install a limited set of attributes in their children, rather than giving them multiple aptitudes to increase their chances of excelling in a variety of activities. Furthermore, many children who are raised by overbearing parents do not seem to lead poor lives. Numerous successful musicians recall how determined their parents were that they practice, many scholars remember being forced to do their homework before they were allowed to go out and play, and what would elite sports be like without parents who dragged their kids to gyms, tennis courts, or ice skating rinks?

And yet, parents may engineer only a limited number of talents in their offspring because of the cost, or for fear that, rather than mastering all, children given a large number of capabilities would master nothing, frittering their lives away as dilettantes or bouncing aimlessly from one activity to the next. There also are bound to be biological limits on how many traits can be enhanced simultaneously, since some desirable characteristics are likely to interfere with others. Children who were engineered to be big and tall, for instance, might not be capable of being lithe; as mentioned earlier, the physically strong might lack fine motor coordination; and an overdeveloped memory may impair the ability to heal from traumatic injury.

Furthermore, we're not just talking about parents charting their children's futures once they were born but *pre*programming them. Until recently, parents could only design their children in very limited ways, such as when actors, athletes, and musicians deliberately marry people with the same talents in order to increase the chances that their children will be able to follow in their footsteps. Interbreeding pales before the ability parents now have to select embryos for implantation after testing their genetic makeup, not to mention the genetic engineering techniques that will be available to parents in the not-so-distant future. Clearly, the more that parents are involved in the process of designing their offspring and the more money and effort they invest in doing so, the less willing they are likely to be to allow their children to march to the beat of their own drums. Even parents who were not particularly overbearing might feel let down by children who did not follow the paths that had been set out for them, and their children might feel guilty for being failures in their parents' eyes.

At the same time, children who did successfully pursue the goals for which they had been designed might feel cheated because they felt that they did not fully deserve the credit they received for their accomplishments. Transhumanist Nick Bostrom dismisses this concern out of hand. "Suppose it turned out that playing Mozart to pregnant mothers improved the child's subsequent musical talent," he speculates. "Nobody would argue for a ban on Mozart-in-the-womb on grounds that we cannot rule out that some psychological woe might befall the child once she discovers that her facility with the violin had been prenatally 'programmed' by her parents. Yet when it comes to e.g. genetic enhancements, arguments that are not so very different from this parody are often put forward as weighty if not conclusive objections by eminent bioconservative writers."[34] Bostrum may be correct that parents already do many things to give their children advantages without seeming to diminish their children's sense of accomplishment, but background music playing in the vicinity of a pregnant woman hardly seems on a par with genetic modification. A recent study seems to bear out the concern: children who were told that they had inherited genes that made them good at sports had lower self-esteem than children who were told that their genes were "neutral" in terms of athletic ability.[35]

While engineered children inherently may resent being deprived of the right to fully enjoy their accomplishments, much of their attitude is likely to depend on how their accomplishments are viewed by others. If they were well rewarded for their attainments despite having had an advantage by being genetically enhanced, children probably would feel more positive about what they had achieved. By the same token, children are likely to take greater pride in their successes the more praise they receive from their parents.

But how would genetically modified children be received by the rest of society? Remember the negative societal reactions to genetically engineered people in novels like *Beggars in Spain*. Opponents of genetic engineering predict that engineered individuals would suffer serious adverse psychological effects from the extremely negative emotional reaction that they would trigger in others. This reaction is known as "the yuck factor," a term attributed to Arthur Caplan as well as Leon Kass, among others. Kass regards this type of revulsion as an indicator of profound moral truths, proclaiming in a famous 1997 article in the *New Republic* that "in crucial cases, repugnance is the emotional expression of deep wisdom, beyond reason's power to fully articulate it."[36] In this case, Kass says that, regardless of the reasoned arguments of the transhumanists, our feeling of repugnance shows us that human evolutionary engineering is simply wrong. Kass's claim that emotional reactions can signal moral truths is highly problematic. After all, as one critic points out, marriages between whites and blacks "turned the stomachs of many in white America,"[37] but this didn't make the marriages immoral. Moreover, emotional reactions to behavior differ among different cultures, which undercuts Kass's conviction that there exist universal moral truths.[38] But regardless of whether emotional responses reveal moral certainties, the yuck factor that engineered children provoked in others could cause them a great deal of suffering.

⌒

One way that children would be especially likely to gross other people out would be if they had been engineered to be part animal and part human. The rationale for inserting animal material into humans is fairly obvious. Julian Savulescu suggests that people may want genes for longevity from tortoises, memory from elephants, and night vision from owls and rabbits.[39] Or imagine if we had an eagle's vision, a dog's sense of smell, a gorilla's strength, a cheetah's speed, and, for those who wanted to spend time underwater, a fish's gills. This is not to say that it will be possible to modify people in these ways anytime soon, or that everyone would want to be given any or all of these abilities, but evolution has produced some excellent traits in other species that at least some parents are bound to be interested in installing in their offspring.

There are various ways that mixing animal and human characteristics could be accomplished. One way would be to create a "chimera," a being that contained both human and animal cells. The NIH, for example, is studying ways to inject human bone marrow into rats to regenerate damaged heart muscles.[40] UCLA researchers have recreated human immune systems in mice to study infections such as HIV,[41] and in order to study neurodegenerative disease, scientists at the Salk Institute in La Jolla have injected human embryonic stem cells into the brains of fetal mice to create mice with working human nerve cells.[42] SUNY Upstate Medical University has a center where mice are injected with human autoimmune disease and cancer cells in order to test possible treatments; disconcertingly perhaps, the technical term they use for what they do to the mice is to "humanize" them.[43] Pigs are also fairly common targets because of the similarities between their organs and those of humans; Chinese scientists have made progress engineering pigs with human cells to reduce the risk of rejection when porcine organs are transplanted into humans.[44] And back in 1992, researchers at the Indiana University School of Medicine and the Veterans Administration in Reno, Nevada, announced that they had produced lambs that had as much as 4 percent of their bone marrow made up of human cells by injecting human blood stem cells into the bellies of the lamb fetuses while they were still in the womb.[45]

Inserting human cells into animals is objected to on the basis that it is unnatural and ugly. Michael Crichton ably evokes this reaction with the character of Dave—part human, part chimpanzee—in *Next*. Critics are also concerned that the animals could become so humanlike as to "introduce inexorable moral confusion in our existing relationships with nonhuman animals and in our future relationships with part-human hybrids and chimeras."[46] The greatest outcry is in response to attempts to grow human brain tissue in the brains of primates. In 2000, for example, a team of researchers led by a Harvard neurologist implanted human neural stem cells into the brains of Old World monkey fetuses.[47] In 2005, an article in the *New York Times* described experiments conducted in St. Kitts in the West Indies in which human neural stem cells were injected into the brains of green vervet monkeys, where they merged with the monkey's brain tissue.[48] The National Academy of Sciences (NAS) reacted to the announcement of these experiments by issuing voluntary guidelines that prohibit permitting an animal into which human embryonic stem cells have been introduced to breed, as well as outlawing the insertion of human embryonic stem cells into animal eggs or sperm and vice versa.[49]

While inserting human cells into animals is now fairly routine, attempts at inserting animal cells into humans is still relatively infrequent. As a result of the shortage of human organs for human transplantation, transplant surgeons have long been interested in replacing faulty human organs with parts from animals, a practice known as xenotransplantation. The first attempt was in 1906, when a French surgeon named Mathieu Jaboulay sewed a pig's kidney into one woman's elbow and a goat's liver into another's. The problem with transplanting animal organs, however, is that the human body regards them as foreign infections and tries to kill or "reject" them. Thus, the pig and goat organs implanted in 1906 failed almost immediately, and both patients died shortly thereafter.[50] More recently, doctors at Loma Linda University Medical Center in 1984 inserted a baboon heart into an infant called Baby Fae, but the child only survived for 20 days. Despite the marginal success thus far, experimentation persists. According to the Biotechnology Industry Organization (BIO), surgeons in the United States and South Africa made 31 attempts between 1963 and 1993 to use xenotranplants as a stopgap measure to keep patients alive until human organs could be procured.[51] One of these patients, who received a chimpanzee kidney, lived for nine months, and in 1992, a woman survived until a human organ was available by relying on a pig's liver, which was kept outside her body and connected to her by arterial tubes.[52] In addition to trying to transplant entire animal organs into people, researchers have attempted to treat Parkinson's disease by inserting brain tissue from fetal pigs into the brains of patients, although the results so far have been disappointing.[53] BIO describes ongoing research in which patients with Huntington's disease are being given modified bovine brain tissue, although the organization admits that "these studies are still very preliminary."[54]

Eventually, scientists may attempt to fold nonhuman DNA into a person's germ line, so that the animal DNA would be passed along to his or her descendants. One technique would be to fertilize human eggs with animal sperm to produce so-called animal-human hybrids. Alternatively, there could be nonhuman-human transgenics, where animal DNA was spliced into human DNA at a sufficiently early stage of embryonic development that the animal DNA found its way to the individual's sperm or eggs.[55] Neither of these techniques is known to be in use as of yet, and with the technology currently available, it is doubtful that any of the resulting embryos would survive until birth.[56] But with sufficient advances in reproductive and genetic science, approaches like these might become options in the future.

Putting animal cells or DNA into humans raises a number of disturbing issues. Safety concerns go far beyond whether the animal material would function as it was intended. Remember how Paul Berg almost came a cropper when he planned to splice a cancer-causing monkey virus into bacteria that populate the human digestive tract. Efforts to implant animal organs or tissue into humans to alleviate transplant shortages have bogged down not only because of rejection problems but because of the fear that the recipients could be infected with lethal animal viruses. (Marburg, ebola, malaria, HIV, and many influenzas are viruses that have migrated from animals to humans.)

Even if the animal material didn't cause humans physical injury, it could cause psychological or social harm to those who had been engineered. So much animal matter might be added to humans that it raised doubts about whether the resulting individuals were still truly human. Philosopher Jonathan Glover observes that "blurring species lines may be upsetting precisely because it breaks down our system of classification," explaining that it could lead to confusion or reduction in respect for resulting individuals.[57] Glover is not worried that the chimeras would be cast out from human society; in his opinion, this particular problem can be solved "by producing enough of the new type for them to make their own community."[58] But what would it be like to be shunned by everyone else? Imagine being treated like the occupants of the "prawn" shantytown in the movie *District 9*.

Even if parents stopped short of making chimeras or animal-human hybrids, children might be considered no longer fully human merely because they had been genetically engineered. Leon Kass, for example, complains that "most of the given bestowals of nature have their given species-specified natures: they are each and all of a given *sort*. Cockroaches and humans are equally bestowed but differently natured. To turn a man into a cockroach—as we don't need Kafka to show us—would be dehumanizing. To try to turn a man into more than a man might be so as well."[59] Of course, as Jonathan Glover notes, there is much about human nature that we could probably do without. "Given our history," he points out, "the idea that we must preserve all the characteristics that are natural to us is not obvious without argument."[60]

Nevertheless, opponents of evolutionary engineering are troubled by the prospect that people who were genetically modified would be denied human dignity. Frances Fukuyama warns, for example, that "denial of the concept of human dignity—that is, of the idea that there is something unique about the human race that entitles every member of the species to a higher moral status than the rest of the natural world—leads us down a very perilous path."[61] Bioethicist and legal scholar George Annas similarly fears that "genetic engineering has the capacity to change the very meaning of what it is to be human. There are limits to how far we can go in changing our human nature," he continues, "without changing our humanity and our basic human values. Because it is the meaning of humanness (our distinctness from other animals) that has given birth to our concepts of both human dignity and human rights, altering our nature necessarily threatens to undermine both human dignity and human rights. With their loss the fundamental belief in human equality would also be lost."[62] When the Council of Europe, one of the two legislative bodies of the European Union, voted to prohibit both "interventions aimed at modifying genetic characteristics not related to a disease or to an ailment" and "interventions seeking to introduce any modification in the genome of any descendants,"[63] it asserted that its objective was "to protect human rights and dignity," which, it added, "is inspired by the principle of the primacy of the human being."[64]

Transhumanists scoff at these concerns. Nick Bostrum observes that the trend in Western societies has been to extend human dignity to new groups. "The set of individuals accorded full moral status by Western societies has actually increased," notes Bostrum, "to include men without property or noble descent, women, and non-white peoples." This leads him to be confident that dignity would be extended to "future posthumans," and perhaps even to "higher primates or human-animal chimaeras."[65]

Moreover, the transhumanists and their opponents disagree passionately about whether humans in fact have a special "nature" that would be jeopardized by evolutionary engineering. Fukuyama maintains that there is a cluster of uniquely human attributes that he calls "Factor X," and he opposes genetic engineering even if it were technically successful on the grounds that it would alter this bundle of traits. Unfortunately, Fukuyama declines to specify exactly what traits he has in mind, claiming that "Factor X cannot be reduced to the possession of moral choice, or reason, or language, or sociability, or sentience, or emotions, or consciousness, or any other quality that has been put forth as a ground for human dignity. It is all of these qualities coming together in a human whole that make up Factor X. Every member of the human species possesses a genetic endowment that allows him or her to become a whole human being, an endowment that distinguishes a human in essence from other types of creatures."[66]

Fukuyama's critics, including many of the transhumanists, argue that he is forced to describe Factor X in this mysterious manner because it is impossible to identify any characteristics or behaviors that are uniquely human. As behavioral biologists have discovered, other animals do virtually everything we do, if perhaps less well. Whales sing. Crows use tools.[67] A species of monkey called Campbell's Guenon speaks, even forming sentences with syntax.[68] The late Swedish archeologist Bo Gräslund notes that "we share most of our emotional repertoire with other higher animals. Anger, fear, anxiety, respect, happiness, the joy of reunion, parental love, sorrow, sadness, loneliness, apathy, friendship and sexual frustration, perhaps also melancholy, longing, jealousy, hate and a lust for revenge are all things that we encounter in some form among several higher animals."[69] My dog loves to play tricks on me, and even if she doesn't giggle, I'm convinced that she grins. The Capuchin monkey has been said to have such a well-developed sense of fairness that, when rewarded unfairly compared to other monkeys for performing the same task (handing a rock to a trainer), it will refuse to perform the task again, reject the reward, or even throw it at the researchers.[70]

One thing we do that animals don't is write, but we only started doing this about 5,500 years ago,[71] and Fukuyama no doubt would accept that humans with full-fledged amounts of Factor X have been around a lot longer than that. The difficulty that Fukuyama and others who share his view of human nature encounter in attempting to specify exactly what separates humans from other animals leads to the accusation that theirs is a discredited, homocentric view of evolution, or what anthropologist Pamela Willoughby calls "old constructions of the past" that "have no more basis in fact than the origin myths of other societies."[72] But even if their views are unsound, parents considering whether to genetically modify their children need to be aware that many people may agree with Fukuyama that genetically engineered individuals may not be worthy of being treated as fully human.

Another reason that evolutionarily engineered children may strike others as abhorrent is the way in which they would be produced. People may be put off by the deliberate, calculated decisions that parents would make about which of their children's characteristics to modify. As the Genetics and Public Policy Center at Johns Hopkins notes, "religious and secular scholars alike have argued that [human germ line genetic engineering] would threaten human dignity by creating children in a utilitarian way."[73] Some people also may be offended by the need to replace the traditional sexual act with artificial insemination, which would be necessary, at least until sometime in the future, in order to isolate and manipulate cells from early-stage embryos. Leon Kass, for example, once expressed the view that IVF was "a degradation of parenthood."[74] Kass has since reconciled himself to the use of IVF as an acceptable means of treating infertility, but he still opposes it for other uses, and he continues to rue it as the technology that opened the door to disturbing technologies such as genetic engineering and human reproductive cloning.

Proponents of evolutionary engineering defend IVF and other reproductive methods that would be required to make it possible by pointing to the critics' insistence that humans are unique. "Since humans are the sole species on earth that can plan and create," argues attorney Joshua Rosenkranz, "perhaps there is something uniquely human about procreation through genetic engineering."[75] Bioethicist and former priest Joseph Fletcher even argues that these new methods of reproduction are *more* human than the traditional methods. "It seems to me," he states, "that laboratory reproduction is radically human compared to conception by ordinary heterosexual intercourse. It is willed, chosen, purposed and controlled, and surely these are among the traits that distinguish Homo sapiens from others in the animal genus, from the primates down. Coital reproduction is, therefore, less human than laboratory reproduction—more fun, to be sure, but with our separation of baby making from love making, both become more human because they are matters of choice, not chance."[76] But traditionalists like Kass are unlikely to be convinced, and the more artificial the reproductive method, the more offended they are likely to be by its results.

It isn't hard to envision the devastating effects that a strong negative reaction from others could have on the self-image and self-esteem of genetically engineered children. Children's psyches are fragile under the best of circumstances; any parent knows how miserable perfectly normal children can become when they are teased for some trivial or even wholly imagined attribute. Now imagine what it would be like for children who truly *were* different, perhaps dramatically so. Moreover, the wounds sustained as a result of being stigmatized and rejected by one's peers have long-lasting effects. Studies show that adverse effects on academic performance persist through young adulthood, and unpopular children are more likely to have serious psychological problems and, especially in the case of boys, to engage in criminal behavior and aggressiveness in later life.[77]

Defenders of evolutionary engineering seek to downplay these concerns. Nick Bostrum reminds us, for example, that "similarly ominous forecasts were made in the seventies about the severe psychological damage that children conceived through IVF would suffer upon learning that they originated from a test tube—a prediction that turned out to be entirely false."[78] Genetically modified children in fact might be treated as celebrities. Leslie and John Brown, the parents of Louise Brown, the first baby conceived through IVF, were forced to go to extraordinary lengths to preserve their privacy when the press discovered when Leslie was six months pregnant what they had been doing.[79] Registered under a false name in a hospital, Leslie and John could not go home because of the hoards of reporters and photographers awaiting them. A British tabloid finally arranged to smuggle them out of the hospital, a feat requiring the use of nine decoy cars. (Surprisingly, Louise herself has been able to lead a normal, retiring life while quietly working for a shipping company in Bristol, England. When upon turning 30 she was contacted by the media and asked how she planned to celebrate her birthday, she replied, "I might go out with my friends or I might have a meal with the family. I'm planning on having a quiet one.")[80]

Whether genetically engineered children would be treated as celebrities or freaks would no doubt depend in large part on the ways in which they were engineered. One of the reasons why Louise Brown has been able to lead a relatively private life is that she looks completely normal. Genetically modified children who did not look normal, however, would find it hard to avoid sticking out, especially if they appeared extremely odd or ugly. Although putting eyes in the back of our heads might be tremendously advantageous, observes British bioethicist John Harris, "this might also make people thus modified so sexually and arguably aesthetically repulsive that it is bad for you, all things considered."[81] Indeed, as we will discuss later, this could be deleterious from an evolutionary standpoint if people with rear-view vision had difficulty finding mates.

Of course, looks are not everything. Even if children's physical appearance was sufficiently normal that it did not attract attention, they could set themselves apart from others by their behavior. A child who was dramatically smarter, faster, stronger, or perpetually cheerful would be bound to be noticed, and not necessarily in a good way. As Ann Hulbert writes in the *New York Times* about the attitudes toward exceptional figures in the past, "baby Hercules had occasion to display his prowess in strangling serpents because jealous Juno, angered that Jupiter had sired a son with a mere mortal, dispatched snakes to his cradle." When spectators saw 12-year-old Jesus sitting in the temple, "in the midst of the doctors, both hearing them, and asking them questions," this invited "not only astonishment 'at his understanding and answers,' but also rebukes from his bewildered parents; they're unsettled by his insistence that he 'must be about my Father's business,' well aware that he isn't referring to Joseph." Hulbert adds that "in the Infancy Gospel of Thomas, perhaps the first early Christian attempt to fill in Jesus' life before that temple story, awe is mixed with terror. Jesus is an alarming little boy who doesn't merely make real birds out of clay and work other miracles but causes the death of those who scold him for not resting on the Sabbath and shames masters who try to instruct him in his letters."[82] Perhaps you recall your reaction to seeing the *Twilight Zone* episode entitled "It's a Good Life," in which a normal-looking boy uses his frightful powers to terrorize the adults around him.

Even if genetically engineered children looked normal and took pains whenever they were being observed to behave the way others did, they might not be able to keep their background a secret for long. Louise Brown's parents reportedly accepted hundreds of thousands of dollars from a British newspaper for the exclusive rights to their story, and if they had rejected the offer, the press and paparazzi most likely would have hunted them down anyway and hounded them for their media value. Even if the families of engineered children sought to maintain their anonymity, opponents of genetic engineering and child welfare advocates might try to sniff them out so that the parents could be punished or at least ostracized for engendering their children.

Added to the physical risks of genetic engineering, then, are the risks to the modified children's mental well-being. They may be socially disadvantaged by parental choices that time proved to be bad or a fad. They may be hurt by family discord provoked by the costs and challenges of raising them or being their siblings. They may suffer the psychological distress of growing up as prodigies, or far worse, as monsters. They may be punished for what they look like, how they act, or simply who they are. The potential adverse impact of genetic engineering on engineered children alone should be enough to make us wary about going down this path.

The physical and emotional impact on engineered children, while the right place to begin our evaluation of evolutionary engineering, is not the only concern, however. We must also consider the potential impact on the societies in which these children will live. Not everyone's children would be genetically modified, at least not at first and probably not for a considerable time. What would happen to social and political institutions if only some of their members were engineered? How would this affect the distribution of societal burdens and benefits? Would the creation of a genetic aristocracy through the use of germ line engineering, even one expected to last only until the genetic technology became more widely available, widen the gulf between haves and have-nots to the point that it fatally frayed the fabric of civil society? And given that living harmoniously alongside people from different cultural backgrounds is proving so difficult in many places, what would it be like to try to coexist with people with major biological differences?

CHAPTER FIVE

Broader Consequences for Society

E ngineered children could suffer considerably if regarded as freaks by those who were not engineered. But there is another factor that would significantly affect how engineered persons were viewed by others: genetic engineering is likely to be unaffordable for large segments of society. As discussed earlier, IVF, which would be a necessary first step in genetically modifying an embryo, costs about $50,000 for each live birth. Since this is the median household income in the United States,[1] it is clear that many Americans (not to mention those living in poorer countries) would lack the resources to pay for IVF, let alone for the genetic engineering. No doubt costs would come down over time, but by then it might be too late.

In Nancy Kress's novel *Beggars in Spain*, only two of the original Sleepless children are born to wealthy parents. This makes sense, explains one of the children, because "rich people don't have their children genetically modified to be superior—they think any offspring of theirs already is superior. . . . We Sleepless are upper-middle class, no more. The children of professors, scientists, people who value brains and time."[2] Kress's view seems highly improbable, however. In a future in which genetic manipulation can confer exceptional abilities, wealthy parents are bound to want to buy their children whatever advantages are being marketed. It is much more likely that *only* families with substantial wealth will be able to engineer their children, at least initially. As I described in my earlier book *Wondergenes*, the families that were able to avail themselves of genetic technologies would be

the most attractive, strongest, most graceful, most intelligent, most charismatic, and most inventive, and they will run the most successful businesses. All of these advantages will be rolled into the same persons. They will enjoy decisive advantages over everyone else in all realms of life—sports and beauty contests, game and talent shows, entertainment and the arts, admission to the best educational institutions, entry into the professions, political office and government appointment, getting rich or richer, and grabbing the most desirable mates. They will attain a monopoly over the best things in life, and their position at the pinnacle of society will be unassailable.[3]

And remember that, as a result of germ line genetic engineering, these advantages would be passed on to their offspring, creating what I call the new "genobility."

Already, our society is being destabilized by the growing rift between rich and poor. As I pointed out in my book *The Price of Perfection*, "corporate executives earn hundreds of times more than their workers, compared to just an 11-to-1 ratio in Japan and a 22-to-1 ratio in Britain. The 300,000 Americans with the highest incomes earned 440 times more than those in the bottom half of the country in terms of income, and their income almost equaled that of the bottom 150 million."[4] In 2006, just over half of household income was concentrated in the top 20 percent of Americans.[5] Census data show that the top 20 percent now own 84 percent of the nation's wealth.[6] George H. Bush joked about this to a wealthy audience in 2000: "What an impressive crowd: the haves, and the have-mores. Some people call you the elite; I call you my base."[7] But it's no joke. A professor at the University of California puts it succinctly: "Just 10% of the people own the United States of America."[8]

What's holding our country together in spite of this growing inequality arguably is a persistent belief that we still enjoy equality of opportunity. In a recent poll, a random sample of 5,500 respondents grossly underestimated wealth disparities in the United States, guessing that the richest 20 percent of the population only owned 58 percent of the wealth, rather than the true figure, 84 percent.[9] We believe that our kids can go to college even if we didn't, and that anyone who works hard can do well. But if the American Dream were exposed as a myth, then liberal democracy would be in danger of collapsing. And nothing is likely to threaten the belief in equality of opportunity as much as inequality engineered into our genes.

As I described in *Wondergenes*, if the oligarchy that already wields disproportionate social and political power obtained preferential access to the advantages that genetic engineering made possible, it might not only discriminate against individuals whom they deem inferior but treat the entire population of these individuals as a pariah caste, like the Dalits or "Untouchables" in India. As the Council for Responsible Genetics warns, "people who fall short of some technically achievable ideal would increasingly be seen as 'damaged goods.'"[10] Emily Marden and Dorothy Nelkin point out the special risks that would be posed for persons with disabilities, who "fear that they will be devalued or be seen as having 'lives not worth living' if it were possible to eradicate their disability."[11] Susannah Baruch and her colleagues at Johns Hopkins's Genetics and Public Policy Center object that "modifying children to eliminate disabilities conflicts with the notion that living with a disability need not be detrimental to that individual, his or her family, or society at large" and caution that "such decisions by prospective parents reduce a disabled person to a single trait and reinforce the idea that the problem is disability, rather than society's failure to provide adequate measures so that those who are disabled can function well."[12]

Yet there is no reason to expect that the genetic underclass would just sit back and allow itself to be subjugated. In my earlier book *Wondergenes*, I predicted that a societal collapse could take place if only wealthy families were able to afford to genetically engineer their children. This breakdown might not occur at first or all at once, since I speculated that

> the genetic underclass might cede power for a time to its genetic superiors in return for the material benefits made possible by genetic advances. The members of the underclass might be content for a while with being upwardly mobile only within the confines of their class. The enhanced, in turn, might rule according to enlightened principles of *noblesse oblige*, taking care to permit sufficient benefits to trickle down to maintain political and social equilibrium. A democracy of sorts even might persist, with the unenhanced electing representatives who either were members of the genetic upper class or who were committed to preserving its privileges. Such a system might not look very different from our own in this respect, since we typically elect representatives who are considerably more privileged than their constituents.

But I doubted that this state of equilibrium could be sustained for long. "For one thing," I noted, "the members of the genetic upper-class would need to maintain a good deal of self-control to avoid overreaching. They would need to monitor and regulate each other to prevent anti-social excesses of greed. And a quasi-democratic system would be highly vulnerable to demagogues who achieved political power by promising to redistribute genetic enhancements more evenly." Ultimately, I stated, "we are likely to encounter an era of growing social chaos as society swings in ever-widening arcs between rule by underclass demagogues and by the genetic aristocracy. Eventually this could deteriorate into mob rule, and finally, anarchy. To rid itself of its underclass status, the unenhanced even might go so far as to try to destroy the scientific foundations of the genetic revolution, physically dismantling research centers and erasing mapping and sequencing data." Boston University professor George Annas agrees that things could get very bad indeed. "The new 'ideal' human, the genetically-engineered 'superior' human, would almost certainly come to represent 'the other,'" says Annas. "If history is a guide, either the normal humans will view the 'better' humans as 'the other,' and seek to control or destroy them, or vice-versa."[13]

Another way that evolutionary engineering could destabilize society would be if it produced too much change too rapidly. As part of their culture, humans have developed a remarkable set of tools. These tools include language, which helps humans to avoid making the same survival-impairing errors over and over again by sharing knowledge and experience and transmitting it easily to their offspring; science, engineering, and industry, including communications technologies that make it possible to transmit language rapidly and richly; machinery to build better dwellings, gain access to additional resources, extend our range, and survive what previously would have been lethal shifts in the environment; and modern medicine, which enables people to reproduce who would not have been able to before.

Over much of time, these instruments of human culture have developed slowly. The first stone tools, which Ian Tattersall describes as "simple flakes chipped from parent 'cores,'" are about 2.5 million years old. Although modern humans appeared around 700,000 years later, on the tool front, nothing much happened for almost another million years, at which point humans invented hand axes. Tattersall explains that it took another million years after that before humans made the next big leap, "prepared-core tools," in which "a stone core was elaborately shaped in such a way that a single blow would detach what was an effectively finished implement."[14] Not exactly a cultural juggernaut.

Fifteen thousand years ago, however, technological change suddenly gathered steam. Humans switched from being hunter-gatherers to farmers and herders. Writing appeared a few thousand years later. In the past 200 years, the industrial and computer revolutions have taken place; in the past 50, the revolution in human genetics. "During this time," points out Swedish geneticist Claus Ramel, "man did not undergo any dramatic genetic alterations, and the whole development of human

society and way of living can be attributed to a cultural and not a genetic evolution."[15] The tools of modern technology arguably now play a
greater role than the forces of nature in making humans better adapted
to the environment. Moreover, reliance on technological fixes may well
become even greater if humans continue to have such limited success
in bringing the environment under their control. Not that we haven't
tried everything from invoking the gods to shooting off "hail cannons,"
sonic blasters that look like upended rocket engines that fruit growers fire in an effort to prevent hail from damaging their crops. For the
2008 Peking Olympics, the Chinese even took the veteran technology of
cloud seeding to a new level, enlisting 37,000 people to fire silver iodide
shells at passing clouds from antiaircraft guns and rocket launchers so
that the clouds would dump their moisture before they reached, and
spoiled the events being held in, the roofless Bird's Nest Stadium. (The
plan also was to use the weapons to make it rain in Peking in case the
atmosphere needed to be cleansed so that it could be breathed by elite
athletes competing at the peak of their performance.) In fact, rather
than making the environment more hospitable and predictable, it seems
that humans are exacerbating the environmental challenges they face
through global warming.

The tools of culture not only protect our vulnerable bodies from environmental threats, including the ones we create, but also can change
the biological makeup of our bodies directly. Even now, before genetic
engineering is perfected, David Sloan Wilson claims that "our capacity
for culture has shifted evolution into hyperspace."[16] Biologist Stephen
Palumbi lists numerous biological changes that have resulted in just the
past 50 years from the effects of human activity on the ecosystem, including "antibiotic resistance, the triumph of HIV over antiviral drugs, size
reduction in overexploited fisheries, and resistance of insects to nerve
gas pesticides."[17] Ernst Mayr points out that the sickle cell trait and other
genes that confer partial resistance to malaria most likely have probably
developed in less than 100 generations.[18] Perhaps the best illustration
of culture impacting human genetics, asserts Nicholas Wade, is lactose
tolerance. Normally, the gene that digests milk switches off after wean-
ing and adults cannot tolerate dairy products. As cattle became domesticated in Northern Europe over the past 5,000 years, the switch appears
to have become inactive in populations that obtain nourishment from
bovine milk.[19]

But what would happen if radical genetic modification caused biological change to take place at an even more rapid pace? In *Wondergenes* I pointed out that we are in the midst of twin scientific revolutions, the revolution in human genetics and the computer revolution. A company headed by Craig Venter deciphered the structural components of the human genome in one-tenth of the time it took the government by utilizing the largest amount of private supercomputing capacity in the world. Think of how much of a difference computerization has made in the world in the past 30 years. Then, to factor in the contribution of genetic science, square it. That's how fast things are moving just in human genetics. It's as if the controlled use of fire and the invention of the wheel had happened in the same half century. Is there a point at which social institutions and relationships would no longer be flexible enough to accommodate this much instability? Would family life and child rearing become dysfunctional? Would political and economic structures fail? Throw into the mix a major environmental change such as global warming, and there could be a complete societal collapse.

One factor that would cause societal strife would be the extreme reactions of those who oppose evolutionary engineering. Genetic modifications that were too sudden or extreme might trigger a backlash against the technology and perhaps against its underlying scientific research base as well. As noted earlier, much of the opposition to evolutionary engineering comes from conservative religious circles. To them, the notion that humans should control their evolution is a sacrilege. Former priest Joseph Fletcher describes those who "feel that mastery drives out mystery, and that when we dig into nature's secrets, so complicated and yet so simple, we lose our sense of awe and humility."[20] In 2000, the report of a two-and-a-half-year project conducted by the AAAS on the concerns raised by germ line genetic engineering concluded that "even if we have the technology to proceed, however, we would need to determine whether IGM [inheritable genetic modification] would offer society a socially, ethically, and *theologically* acceptable alternative to other technologies" (emphasis added).[21] Transhumanists are dismayed, for example, by polls showing that 80 percent of Americans believe in miracles, one-third claim they receive answers to their prayer requests at least once a month, half the people accept creationism, a majority say that angels and demons are active in the world, and one in five Christians speaks or prays in tongues.[22]

Nonbelievers may be tempted to dismiss these religious concerns. Fletcher himself states, for example, that "the future is not to be sought in the stars but in us, in human beings—because that is where our needs lie. There are no 'acts of God' anymore."[23] Robert Franceour wonders what makes opponents of evolutionary engineering so confident that they know God's wishes. As far as he is concerned, "creation is our God-given role, and our task is the ongoing creation of the yet unfinished, still evolving nature of man."[24] While acknowledging that "playing God is playing with fire," Dworkin points out that "that is what we mortals have done since Prometheus, the patron saint of dangerous discovery. We play with fire and take the consequences, because the alternative is cowardice in the face of the unknown."[25]

It would be foolhardy, however, to disregard the political strength and determination of opponents of evolutionary engineering. In part owing to their pressure, the NIH currently will not fund any research projects involving intentional human germ line genetic modification or human biomedical enhancement. During his two terms in office, President George W. Bush hindered research on human embryonic stem cells and established a Council on Bioethics, headed during most of its tenure by Kass, who used it in large part to expound the views of the religious Right, including their opposition to directed human evolution.

A major battleground in the conflict between science and the religious Right has been the public schools. In 1987, the U.S. Supreme Court held unconstitutional a Louisiana law requiring that public schools teach "creation science."[26] Fundamentalist Christians had devised creation science in an attempt to use scientific evidence to refute the theory of evolution, which they hoped would get around earlier rulings by the court that laws prohibiting the teaching of evolution were unconstitutional. The court was not fooled by creation science; a majority of the justices recognized that it was a religious wolf in science's clothing. In a dissent, however, Justice Antonin Scalia claimed that there was "ample uncontradicted testimony that 'creation science' is a body of scientific knowledge rather than revealed belief" and "infinitely less" reason to hold that "scientific evidence for evolution is so conclusive that no one could be gullible enough to believe that there is any real scientific evidence to the contrary."[27] Taking heart from Scalia, members of the school board in Dover, Pennsylvania, notified high school teachers in 2004 that henceforth they would be required to read a statement to their students declaring, among other things, that "because Darwin's Theory is a theory, it continues to be tested as new evidence is discovered. The Theory is not a fact. Gaps in the Theory exist for which there is no evidence. A theory is defined as a well-tested explanation that unifies a broad range of observations."[28]

Since creation science had been declared a taboo subject in public schools, the fundamentalists also came up with a new approach they called "intelligent design," described by them as a scientific theory that says that the natural world is too complex to be the product of natural evolution. To distinguish intelligent design from earlier attempts to introduce creationist beliefs into public education, the intelligent design proponents claim that in contrast to creationism, which "typically starts with a religious text and tries to see how the findings of science can be reconciled to it . . . , intelligent design starts with the empirical evidence of nature and seeks to ascertain what inferences can be drawn from that evidence." Moreover, "unlike creationism, the scientific theory of intelligent design does not claim that modern biology can identify whether the intelligent cause detected through science is supernatural."[29] The Dover School Board dutifully planned to require students to be instructed that "Intelligent Design is an explanation of the origin of life that differs from Darwin's view" and that, as "with respect to any theory, students are encouraged to keep an open mind. The school leaves the discussion of the Origins of Life to individual students and their families."[30]

The efforts of the intelligent design advocates and the Dover School Board to pretend that they were not trying to substitute religion for science didn't fool U.S. District Judge John E. Jones, who somewhat ironically is a Republican who was appointed to the federal bench by George W. Bush. In a suit brought against the school board by a number of parents, Judge Jones described the school board's policy as "breathtaking inanity" and struck it down as a violation of the First Amendment's separation of church and state.[31] (This earned him the opprobrium of, among others, conservative gadfly Phyllis Schlafly, who called him "biased" and "religiously bigoted," adding, "Judge John E. Jones III could still be chairman of the Pennsylvania Liquor Control Board if millions of evangelical Christians had not pulled the lever for George W. Bush in 2000. Yet this federal judge, who owes his position entirely to those voters and the president who appointed him, stuck the knife in the backs of those who brought him to the dance.")[32]

The creationists have not given up, however. As recently as 2005, Florida, Kentucky, Mississippi, and Oklahoma did not mention evolution in their science curriculum standards, and laws challenging the teaching of evolution were pending or being proposed in 20 states, including Michigan and New York.[33] And in 2008, the Louisiana legislature tried again, enacting the "Louisiana Science Education Act," which ordered the state to "allow and assist" teachers "to create and foster an environment within public elementary and secondary schools that promotes critical thinking skills, logical analysis, and open and objective discussion of scientific theories being studied including, but not limited to, evolution, the origins of life, global warming, and human cloning."[34] Only three Louisiana legislators in the State House of Representatives voted against the bill; in the State Senate, the vote in favor was unanimous. An editorial in the Baton-Rouge *Times-Picayune* called the law "straight out of the Dark Ages."[35]

Creationists and their allies can be counted on to oppose any use of genetic engineering that has evolutionary implications. But extreme forms of genetic engineering could be so repellant that people who ordinarily were receptive to scientific advances might be persuaded to join the creationists' crusade. Reproductive technologies already are shocking to much of the public. Look at pictures of pregnant transgender men such as Thomas Beatie and Scott Moore Biel. As Guy Trebay writes in an article in the *New York Times* entitled "He's Pregnant. You're Speechless," Beatie is "partly a carnival sideshow and partly a glimpse at shifting sexual tectonics, and his image and story powered past traditional definitions of gender and exposed a realm that seemed more than passing strange to some observers."[36]

In the course I teach on genetics and the law, the students read a 1998 California case called *In re Marriage of Buzzanca.*[37] Luanne and John Buzzanca fertilized a donor egg with a donor sperm and had a surrogate mother carry it to term for them. When the couple split up, the question arose of who the child's legal parents were, since no one wanted to be responsible for paying child support. Owing to the manner of the child's birth, the trial judge reached the bizarre conclusion that, from a legal standpoint, *the child had no parents at all.* (Fortunately for the child, the appellate court disagreed and declared Luanne and John to be her lawful parents.)

Now imagine the response if somebody jumped the gun and produced a human who was visibly odd looking, for example, a human-animal hybrid. Chapter 4 focused on the harm that children could suffer if parents modified them in such disturbing ways. But the "yuck" reaction that this type of alteration might produce in others could provoke widespread hostility not only toward the children but toward genetic engineering technologies and those who researched and provided them. There might even be violence against researchers, physicians, and parents and violent confrontations between pro- and anti-engineering forces. It is not hard to imagine anti-engineering groups launching a terrorist campaign similar to that waged by certain pro-life groups and individuals against abortion clinics and doctors. According to the National Abortion Federation, in the United States alone, pro-life forces have been responsible for 8 murders, 17 attempted murders, 4 kidnappings, 41 bombings, 175 incidents of arson, 97 cases of attempted bombings or arson, 184 cases of assault and battery, 100 acid attacks, 661 anthrax threats, 416 death threats, and 1,429 incidents of vandalism.[38]

The violence against geneticists might come not only from religious extremists, moreover, but from radical environmentalists opposed to genetically modified foods, or just plain sickos. Protesting technology, including genetic engineering, Ted Kaczynski, the so-called Unabomber, set off 16 bombs over 23 years, killing three people and seriously wounding 23. In his manifesto *Industrial Society and the Future*, in which he explained the motivation for the bombings, Kaczynski gave the following description of what he thought of gene therapy:

> Suppose for example that a cure for diabetes is discovered. People with a genetic tendency to diabetes will then be able to survive and reproduce as well as anyone else. Natural selection against genes for diabetes will cease and such genes will spread throughout the population. (This may be occurring to some extent already, since diabetes, while not curable, can be controlled through the use of insulin.) The same thing will happen with many other diseases susceptibility to which is affected by genetic degradation of the population [*sic*]. The only solution will be some sort of eugenics program or extensive genetic engineering of human beings, so that man in the future will no longer be a creation of nature, or of chance, or of God (depending on your religious or philosophical opinions), but a manufactured product. . . . If you think that big government interferes in your life too much *now*, just wait till the government starts regulating the genetic constitution of your children. Such regulation will inevitably follow the introduction of genetic engineering of human beings, because the consequences of unregulated genetic engineering would be disastrous.[39]

A coalition of Christian conservatives and radical Neo-Luddites could seriously disrupt scientific progress in general and advances in genetic engineering in particular. They might gain sufficient political power to require scientists to pass an ideological litmus test in order to receive government research funding or to be allowed to teach at public universities. They might hold antiscience witch hunts like the Inquisition's censure and punishment of Galileo for his heliocentric views. Liberals and scientists would push back. Things could get ugly.

But evolutionary engineering could do more than merely create warring social classes and ideological factions. It could create separate human subspecies. It is well known that people tend to mate with people who are like themselves,[40] a phenomenon known as "positive assortative mating" or "homogamy." Although we previously identified a number of reasons why the bulk of society, unable to afford genetic engineering or rejecting it on ethical or religious grounds, might ostracize a smaller number of genetically engineered children, it is just as likely that the genobility would refuse to interact socially with those whom they felt were their inferiors and, in particular, would decline to reproduce with them.

Evolutionary researcher Oliver Curry created somewhat of a stir when he forecast such a future in a 2006 essay for the British television channel Bravo. Curry claims that the essay was supposed to be "a 'science fiction' way of illustrating some aspects of evolutionary theory,"[41] but it was picked up by the British press and portrayed as scientific prediction. Curry's essay envisioned that by 3000, humans will have attained "the peak of human enhancement." They will live until they are 120 and display features that are constantly and universally found attractive. Women will have "a 0.7 waste-to-hip ratio, lighter-than average skin colour, smooth hairless skin, glossy hair, symmetry, large clear eyes, low testosterone (e.g., small chin), [and] pert breasts," while males will have "a 0.9 waste-to-hip ratio, [be] taller-than-average [a reference to the improbable demographics of Lake Wobegon?], [display] symmetry, cues of high testosterone (for example, square jaw, deep voice, bigger penis), and athleticism." Curry's vision of future males led Britain's largest selling newspaper, the *Sun*, to carry the story of his report under the headline "Good News. All Men Will Have Big Willies. Bad News. It Won't Happen till Year 3000."[42]

Curry's timeline didn't stop at the year 3000. By the year 100,000, he imagined that mating between "extreme" individuals would create such genetic, social, and economic inequality that "the circles in which the genetic elites move become ever more exclusive, until they lose contact altogether with the rest of society, and come to constitute their own 'celebrity' gene pool." This will give rise to two human subspecies: "People with the best genes will have chosen to mate with each other, leaving the rest to mate amongst themselves." Borrowing names paleontologists use to describe two different types of prehuman *Australopithecines*, Curry calls the descendants of the genetic upper class "gracile" (meaning graceful) and the offspring of the lower class "robust." Eventually, he says, they will come to resemble the "elfin" Eloi and "brutish" Morlocks of H. G. Wells's *The Time Machine*: "The genetic 'haves' will tend to be tall, thin, symmetrical, clean, healthy, intelligent and creative. The genetic 'have nots' will be short, stocky, asymmetrical, grubby, unhealthy and less intelligent." True, Curry is describing things 100,000 years down the road and H. G. Wells's time traveler doesn't encounter the Eloi and Morlocks until the year 802,701, but with advances in genetic engineering and a society increasingly polarized between rich and poor, who can say whether the creation of dominant and subordinated human subspecies would actually take that long.

If the separation between engineered and nonengineered humans became great enough, the result could even be the creation of entirely separate humanoid species. The process by which one species succeeds another is complicated. A "daughter" species can split off (called "speciation") and survive while the parent species becomes extinct, or a species can change over time to such an extent that it forms a new species (called "anagenesis" or "phyletic evolution"). In either case, a widely accepted defining characteristic of a species that reproduces sexually is that its members can only breed with each other and not with other species. Thus, separate species are said to be "reproductively isolated," or as biologist H. Allen Orr explains, "they have genetically based traits preventing them from exchanging genes."[43] If Neanderthals were a separate species from *Homo sapiens sapiens*, for example, then the two species would not have been able to interbreed; if they could interbreed, then they were merely different subspecies of the same species. In order for evolutionary engineering to create a new human species, then, something has to happen to make its members incapable of reproducing with members of the old.

In traditional evolution, the impetus for speciation often has been geographic isolation. The classic example is the finches that Charles Darwin discovered on the Galapagos Islands when he visited them during his famous voyage on the *Beagle*. The beaks of birds on different islands were so unlike that Darwin mistakenly thought that they belonged not just to different species but to different families (a broader biological classification). Only after he returned to England did he determine that the distance between the islands had merely led the birds to form separate species.

Geographic isolation is largely a thing of the past for modern humans. Ernst Mayr asserts that the probability that we will break up into several species is "none at all" and points out that "humans occupy all the conceivable niches from the Arctic to the tropics that a humanlike animal might occupy. Furthermore, there is no geographic isolation between any of the human populations."[44] Technological advances such as ships and airplanes have enabled humans to overcome any physical barriers to interbreeding, and people increasingly are on the move. The United Nations estimates that there are now 214 million migrants worldwide, up almost 40 percent from 1990.[45] Currently one-quarter of Americans under 18 are immigrants or the children of immigrants. The mobility of our species is so great that evolutionary biologists have largely rejected the notion that natural evolution would be able to produce more than one terrestrial human species, and paleontologist Peter Ward believes that geographic isolation will only resume operation when we begin to colonize other planets: "As long as humans are restricted to the surface of the earth, such an event seems unlikely."[46]

Natural evolutionary forces may be incapable of creating a new human species because of geographic isolation, but what about the unnatural evolution that would result from genetic engineering? Gregory Stock thinks that the lack of geographic barriers will prevent new species from forming even with evolutionary engineering. "Reproductive isolation is central to speciation," he observes, and "such isolation is unlikely to occur in future human subpopulations." Even space travel may not change this, he argues: "Not only will our offspring remain in close physical proximity, unless and until humans migrate out into the vast seas of space, but genomics and advanced reproductive technologies are breaching the barriers to genetic exchange even among different species. . . . If scientists in Oregon can already give a jellyfish gene to a primate, surely we will continue to be able to exchange genes with one another."[47] Transhumanist Nick Bostrom agrees that speciation is unlikely: "The assumption that inheritable genetic modifications or other human enhancement technologies would lead to two distinct and separate species should also be questioned. It seems much more likely that there would be a continuum of differently modified or enhanced individuals, which would overlap with the continuum of as-yet unenhanced humans."[48]

But geographic isolation isn't the only reason that groups stop interbreeding. Another is "mechanical isolation," a fancy way of saying that a male's round peg won't fit into a female's square hole. What would lead parents to engineer their children so that their sexual organs were compatible only with other, similarly engineered individuals? Surprisingly perhaps, transhumanists don't pay much attention to the sexual behavior or equipment of future humans beyond suggesting that gender and sexual preference will become blurrier and sex toys will morph into sex partners. Science fiction also is strikingly unimaginative about the evolution of the human sexual apparatus. Theodore Sturgeon envisioned societies of androgynous hermaphrodites in his 1960 novel *Venus Plus X*, while Storm Constantine's 1991 novel *Hermetech* features a character who has been altered surgically, but not genetically, to have multiple sexual orifices. A lot of future scenarios rather depressingly forecast that sex and sexual reproduction will disappear altogether, for instance, Michel Houellebecq's *The Possibility of an Island*, where they have been replaced by cloning.

Oliver Curry, who wrote the essay for Bravo television mentioned earlier in this chapter, is one of the few futurists who predicts a change in sexual equipment, namely, the enlargement of the penis. Penis size and prominence play an important role in human evolutionary theory. Anthropologist Nancy M. Tanner, for example, believes that large penises were selected for when humans began to walk upright, which made the organs more visible to females, particularly, Tanner adds, when the organs themselves were "upright."[49] (Her fellow anthropologist Dean Falk calls this the "flasher" theory, which also accounts, he says, for the development of large breasts in females.)[50] If Curry is right, could penises become so enlarged that female vaginas would have to be reengineered to accommodate them? If so, then this could be a sufficient hindrance to reproduction between modified and unmodified individuals that they might be able to evolve into different species.

If future humans were engineered to be smarter, moreover, the size of their heads could increase. The previous chapter, which mentioned this as a potentially lethal complication of childbirth for genetically modified infants and their mothers, conceded that the risk might be reduced by engineering females to have wider pelvises. But then only these women would be able to bear these fat-headed children without serious risk. This would separate them reproductively from other child bearers, which over time could lead them and their mates to also form a separate species.

The most likely way that a new human species would emerge, however, would be through a third type of reproductive isolation: behavioral. Different species display differences in behavior, such as distinctive mating rituals, that identify males and females of the same species to one another. For example, Dutch ornithologist A. C. Perdeck identified differences in the notes of the courtship "songs" of two species of grasshopper that lived in the same habitats, and he demonstrated by the birth of hybrids in the laboratory that they were physically capable of interbreeding.[51] Mating rituals certainly play a role in maintaining ethnic separation among humans; a delightful film portrayal is in the *Godfather*, when Al Pacino's character Michael Corleone, hiding out in Sicily, has to abandon his brusque American style and adopt traditional Sicilian manners in order to obtain permission from a father to court his daughter.

But the behaviors that produce behavioral isolation are not limited to courtship rituals. They include deliberate decisions to refrain from mating with certain types of individuals. In the case of genetically engineered people, behavioral reproductive isolation could occur because they were ostracized by the rest of society like the Sleepless in *Beggars in Spain*, or it could be self-imposed in order to maintain their privileged status in the same way that aristocracies preserve themselves by refraining from intermarriage with commoners, like the Optimen in Heinlein's *The Eyes of Heisenberg*. Thus, Gregory Stock's claim that genetic engineering will enable humans to overcome even the geographic impediments to mating imposed by space travel overlooks the very real possibility that genetic haves simply may not want to breed with have-nots, or that "normal" people may not want to reproduce with those whom they regard as genetic freaks. Peter Ward describes just such a scenario:

> Some parents allow their unborn children to be genetically altered to enhance their intelligence, looks, and longevity. Let's assume that these children are as smart as they are long-lived—they have IQs of 150, and a maximum age of 150 as well. Unlike us, these new humans can breed for eighty years or more. Thus they have more children—and because they are both smart and live a long time, they accumulate wealth in ways different from us. Very quickly there will be pressure on these new humans to breed with others of their kind. Just as quickly, they will be-

come behavioral outcasts. With some sort of presumably self-imposed geographic or social segregation, genetic drift might occur and, given enough time, might allow the differentiation of these forms into a new human species.[52]

This brings us to an important point. From the standpoint of evolution, there is nothing wrong with the creation of a new human species. Evolution is all about species disappearing and new species appearing. Modern humans, for example, are descended from a long line of species that no longer exist. As the next chapter shows, the exact trajectory of human evolution is still a matter of intense debate, but there is general agreement that before (and possibly alongside) *Homo sapiens* there were a number of other species of the genus *Homo*, preceded by species belonging to other hominin genera, which followed other hominid species, which in turn descended from other primate species. Of this long line of species, only one, *Homo sapiens*, is still around.

In fact, the transhumanist vision of a future in which evolutionary engineering creates a more advanced species of *Homo*, or, for that matter, a species belonging to an entirely new hominid genus, is perfectly consistent with the succession of species in the past. Over the course of the human journey, when one species disappeared, another took its place and carried ancestral genetic material onward. There was once a hominid species called *Homo ergaster*, for example. The last member of this species died out some 1.5 million years ago, but it gave rise to other species that eventually became *Homo sapiens*. In short, so long as there is a successor species that carries on the human lineage, there should be no cause for alarm.

Fair enough, but what if more than one of the diverging human species were to exist *at the same time*? Were evolutionary engineering to create a new human species, let's call it "changed human" or *Homo mutatus*, then there would likely be a period during which the members of the new species lived at the same time as *Homo sapiens*. During this period, how well would they get along? Previously we discussed the social and psychological harm that genetically engineered children might suffer if they appeared different than their peers or parents, as well as the stress on social and political institutions if more and less genetically advanced populations or subspecies vied for political power and scarce societal resources. But what would the social, psychological, and political consequences be if the two groups stopped mating together long enough to actually form two separate species?

It may come as a surprise to many that there have been multiple human species in existence simultaneously during the recent past, indeed, on the scale of evolutionary history, virtually in our own time. Only 50,000 years ago, when modern humans inhabited northeast Africa, members of an ancestral human species, *Homo erectus*, could still be found in Asia.[53] And as recently as 18,000 years ago, a stone-tool-making, fire-wielding human species called *Homo floresiensis*, dubbed "hobbits" because of their diminutive size, occupied an island off the coast of Indonesia and produced their cultural achievements despite having the brain size of a chimpanzee.[54] It is not certain whether the hobbits, as the dig workers dubbed these doll-like humans, had a chance to encounter modern humans before they became extinct, and not much more is known about the fate of the *Homo erectus* population in Asia. But we know that there was one other type of human some 30,000 to 50,000 years ago that did come into contact with a different type of human, and not long afterward, it disappeared. Why? What happened to the Neanderthals?

The short answer is that no one knows, and making sense of the available evidence turns out to be one of the great puzzles in modern science. Until May 2010, in fact, many experts thought that the Neanderthals were an entirely separate species from *Homo sapiens*. But then a group of researchers sequenced the Neanderthal genome and, when they compared it to DNA from modern-day humans, found that between 1 and 4 percent of modern human DNA came from the Neanderthals.[55] Among other things, this showed that Neanderthals had mated successfully with *Homo sapiens*, which means that, according to the generally accepted definition of a species, they were members of the same one, just in a different subspecies. (So since modern humans are *Homo sapiens sapiens*, Neanderthals should now be called *Homo sapiens neanderthalensis*, rather than *Homo neanderthalensis*.)

When most people think of Neanderthals, the image that comes to mind is likely to be the classic "caveman," similar to the creatures that Jean Auel called "the Clan" in *The Clan of the Cave Bear* and her other novels. This popular conception has fueled the common belief that the Neanderthals became extinct because they were inferior to and could not compete successfully with the smarter and more skillful modern humans. Noting the absence of domestic artifacts like cooking utensils and the heartiness of female bones, for example, anthropologists Steven Kuhn and Mary Stiner think that Neanderthals died out because both women and men hunted for large game, whereas modern humans had developed a more efficient division of labor between males and females, with the males hunting and the females taking care of the crops and kids at home.[56] Stanford professor Richard Klein similarly argues that modern humans had more advanced language skills and cognitive abilities that enabled them to be better adapted to their environment.[57]

Yet it is now generally agreed that Neanderthals were not stupid; in fact, their brains were larger than those of the modern humans who entered their territory, including ours.[58] Moreover, it is hard to accept the notion that modern humans, having come from the more tropical climate of Africa as most experts believe, would have been better suited to the harsher conditions they found in Europe than the indigenous Neanderthal population that had been living there for 200,000 years. Critics of the inferiority theory also insist that Neanderthals had an advanced culture, based on excavations of what appear to be ritualistic Neanderthal burial sites and the modern-looking objects that were found in them.[59] But supporters of the inferiority theory are not convinced. Thomas Wynn and Frederick Coolidge at the University of Colorado argue that brain shape is more important than brain size, and that the shape of modern human brains made them smarter.[60] As for the burial sites and artifacts, some experts claim that what look like ritualistic Neanderthal burials were merely efforts to conceal their dead so that they wouldn't attract dangerous predators to their settlements, and that either the artifacts found at the burial sites are not really modern looking or the Neanderthals obtained them by trade or theft from their true makers, the modern humans who lived nearby.[61]

So why did the Neanderthals disappear? Another theory is that they succumbed to sudden climate change,[62] but critics dispute that there were any significant climate shifts at the time.[63] Some experts think that the Neanderthals may have died of diseases that they caught from modern humans, like the infections brought by Europeans that devastated Native Americans, while others think that the Neanderthals became cannibals and infected themselves with spongiform illnesses similar to the ones that humans can catch from diseased cows. Still other theories propose that cave living exposed the Neanderthals to smoke and other toxic chemicals that rendered them sterile.[64] And the recent discovery that

Neanderthals and modern humans reproduced with each other provides support for yet one more theory, that the Neanderthals simply interbred themselves out of existence. In other words, they mated often enough with modern humans that their genome was eventually absorbed and submerged, leaving one subspecies, still labeled modern humans, but containing DNA in its gene pool from its Neanderthal cousins.

But there is one more possibility that cannot be ruled out. Rather than interacting and reproducing harmoniously with the Neanderthals, modern humans may have killed them off. This certainly represents a common interaction between different cultures throughout human history. After studying the DNA and behavior of the different groups of tribal hunter-gatherers that still exist today, for instance, Spanish molecular biologist Eduardo Moreno claims that the modern humans whom the Neanderthals encountered were especially violent and warlike.[65] In 2009, he created a stir when he announced that he had found a Neanderthal jawbone amid the bones of modern humans and that the jawbone bore marks on it showing that it had been butchered in the same manner that Stone Age humans cut up deer carcasses.[66] According to Moreno, this proved that the modern humans were cannibals and that the Neanderthals vanished because they were eaten.

If the Neanderthals became extinct because they were hunted down and exterminated by the modern humans who invaded their habitats, to say nothing of being consumed as food, this does not bode well for a future in which normal and genetically modified humans sought to share the same planet, particularly if, as no longer appears to be the case with the Neanderthals, the humans of the future diverged into completely different species rather than just into different subspecies.[67] In an article he wrote in 2000 entitled "The Man on the Moon, Immortality, and other Millennial Myths: The Prospects and Perils of Human Genetic Engineering," George Annas predicted just such a fate: "Ultimately, it almost seems inevitable that genetic engineering would move Homo sapiens to develop into two separable species: the standard-issue human beings would become like the 'savages' of the pre-Columbian Americas, and be seen by the new, genetically-enhanced neo-humans as heathens who can properly be slaughtered and subjugated."[68] And although, as we saw earlier, the famed evolutionary biologist Ernst Mayr believed that the chances of humans breaking up into more than one species were nil, he acknowledged that, if this were to happen, it wouldn't be pretty, since according to the evolutionary principle of "competitive exclusion," multiple human species would not be able to coexist if they required similar things to survive, lived near each other, and one was superior.[69] The creation of competing engineered and nonengineered humans doesn't faze Oxford transhumanist Nick Bostrom. "Human society," he says, "is always at risk of some group deciding to view another group of humans as fit for slavery or slaughter. To counteract such tendencies, modern societies have created laws and institutions, and endowed them with powers of enforcement, that act to prevent groups of citizens from enslaving or slaughtering one another." Nor does Bostrom think that the situation would get out of control if one group was better adapted from an evolutionary standpoint than the other: "The efficacy of these [societal] institutions does not depend on all citizens having equal capacities. Modern, peaceful societies can have large numbers of people with diminished physical or mental capacities along with many other people who may be exceptionally physically strong or healthy or intellectually talented in various ways. Adding people with technologically enhanced capacities to this already broad distribution of ability would not need to rip society apart or trigger genocide or enslavement."[70]

But even assuming that Bostrom is correct about the prospects for peaceful cohabitation between members of the same species who were normal and who were genetically enhanced, which, given the present state of the world, may well be doubted, this doesn't say that two different humanoid species could survive in each others' presence. At the very least, the extinction of the Neanderthals must give proponents of evolutionary engineering pause. Moreover, the technology at the time of modern humans and Neanderthals was primitive. Think what might happen if two warring species went at each other with weapons of mass destruction. The replacement of *Homo sapiens* by *Homo mutatus* might not be problematic, at least from an evolutionary standpoint, but what if conflict between them destroyed *both* species? For that matter, what if aggressive genetic engineering did not result in the creation of a new human species, but instead caused the extinction of the human evolutionary *lineage*, the entire line of human descent?

When a lineage disappears, as opposed merely to a species, the entire genetic load that it has carried since the beginning of life becomes extinct. This is obviously undesirable from the point of view of the genes, and therefore it would stand to reason that humans would be genetically hardwired to regard the loss of their lineage as a catastrophe to be avoided at all costs. But the complete destruction of the human lineage means more than just the loss of a vehicle to transport genes efficiently from one generation to the next. It would seem to confirm Richard Dawkins's bleak words at the beginning of the book, that "the universe that we observe has precisely the properties we should expect if there is, at bottom, no design, no purpose, no evil, no good, nothing but pitiless indifference."[71] The end of our lineage would make our existence pointless, unless you are content with the notion that cultural progress, accumulated human knowledge, and all that we, our ancestors, and our descendants had labored to create could be transferrable to some other form of life, perhaps an intelligent tomato or the machine life that Ray Kurzweil predicts will take over from us in his 2005 book *The Singularity Is Near*. Otherwise, the annihilation of the human lineage would be, to say the least, a shame. As Jonathan Glover puts it, we must not "let the 'conversation of mankind' fade out."[72]

Indeed, it can be argued that it is precisely how discouraging the loss of our lineage seems to us that truly separates humans from other animals. Within the animal world, it is possible to find some manifestation of every aspect of human culture, albeit at a much less sophisticated level—every aspect, that is, except one: no other animal appears to care where its species is headed. In the end, perhaps this determination to safeguard the future of our lineage is what makes humans unique, the elusive endowment that Frances Fukuyama calls Factor X.

How then do we safeguard the future of humanity? Let us look more closely at what aspects of evolutionary engineering might threaten our lineage.

CHAPTER SIX

The End of the Human Lineage

Should society be reduced to a division between genetic haves and have-nots, the human species, along with other life on the planet, could be completely eliminated. Not all of the reasons that a lineage can fail are potential effects of evolutionary engineering. Lineages, like species, can die out because they lose their habitat.[1] If the places where humans can live became uninhabitable, then humans would disappear unless they were able to evolve very quickly and adapt to the new environment. Another cause of extinction that is not directly related to genetic engineering is genetic drift, where the genetic characteristics of an evolutionary line change randomly over time in such a way that the line cannot survive when its environment changes. Sudden, catastrophic environmental change is a third and perhaps most familiar type of extinction event, from the destruction of habitat or from other sources, such as a major disruption of the food chain.

Other harm that might threaten the continuation of the lineage can be caused by evolutionary engineering. As the experience with genetically modified animals and plants demonstrates, harm to future generations may occur in ways that may be hard to foresee. The experience with the hormone diethylstilbestrol (DES), albeit not involving intentional genetic engineering, serves as an example in humans. Between 1938 and 1971, doctors prescribed DES to many pregnant women to prevent miscarriages and premature labor. It was known that the drug gave the women a modest increased risk of breast cancer, but what was not anticipated was that their daughters, upon reaching puberty, would have a marked risk for clear cell adenocarcinoma, a rare form of cervical cancer.[2] A serious side effect, in other words, had skipped a generation. Although the mechanism that causes this cancer is not certain, there is speculation that it is epigenetic, that is, that DES adversely affects the female reproductive tract during fetal development by disrupting the regulatory regions of DNA.[3] If the unforeseen consequences of intentional evolutionary engineering were sufficiently dire, and if they affected a large enough number of people, the harm could extend beyond individuals to imperil whole populations, entire generations, or even the human lineage itself. The Council for Responsible Genetics warns that "inserting new segments of DNA into the germ line could have major, unpredictable consequences for both the individual and *the future of the species*" (emphasis added),[4] in other words, threatening not just the current version of humans but our future embodiment.

Even if genetic engineering did not actually make future persons ill, it could place them at a biological disadvantage in the face of abrupt shifts in the environment. For example, parents might opt for darker children with lots of melanin in their skin cells to protect them against solar radiation that a depleted ozone layer would allow to penetrate the atmosphere. But a large-scale volcanic eruption could block out enough sunlight that lighter-skinned people proved better able to survive, since lighter skin enables the body to manufacture more vitamin D from less sunlight. A deficiency in natural vitamin D might not be a serious problem if people could make do with a synthetic version, but not all genetic changes might be able to cope so easily with sudden environmental changes.

A major concern, for example, is excessive genetic homogenization. A species with a diverse gene pool, where lots of individuals have lots of differences in their DNA, is more likely to survive an unexpected environmental threat than a species that shares more of the same genetic variations, since there is a greater chance that some of its members will possess a genetic makeup that affords them protection. Environmental challenges may stem from planetary phenomena or may be extraterrestrial. Recall Michael Crichton's *Andromeda Strain,* in which a microbe brought back to earth from space was deadly to all but those who had an abnormal pH level in their blood. If for some reason genetic engineering made everyone's pH level normal, no one would be able to survive the infection. Crichton's novel was science fiction, but in 2003, the prestigious British medical journal the *Lancet* published a letter from a group of British and Indian scientists proposing that both the 1918-19 influenza virus, which killed 50 million people worldwide, and the SARS virus, which in 2003 killed almost 800 people and provoked public hysteria throughout Asia, came from outer space.[5]

Excessive genetic homogenization has already caused catastrophes in agriculture. Wheat is susceptible to a fungus called stem rust. U.S. growers plant wheat that has been bred to be genetically resistant to the common types of the fungus, but this leaves them open to infection by rarer types. In the early 1950s, this vulnerability caused the destruction of as much as half of the wheat crop in some states.[6] In 1970 a fungus blight attacked the American corn crop. What happened is complicated, so bear with me. Among major crops, corn is unusual: its plants have separate male parts, the tassles, which contain pollen, and female parts, the ears, which contain the seeds, in the form of incipient kernels. In order for the kernels to form completely, they must be pollinated. Left alone, a corn plant will pollinate itself. But since the 1930s, farmers have been aware that hybrid corn, that is, corn that is a blend of the genes of different strains, is healthier and yields a bigger crop. To produce hybrid corn in the past, farmers had to expose the seeds to pollen from different strains, which meant preventing the plant from fertilizing itself. This in turn required them to remove the male tassels from the plant, a laborious process only part of which could be accomplished by machine. In the mid-1960s, a strain of corn was discovered that was sterile, that is, producing no pollen, so it could be hybridized without being detasseled. By 1970, 85 percent of the corn grown in the United States contained genes from this sterile corn. So far, so good. But it turned out that this sterile corn was especially susceptible to a type of fungus called Southern corn leaf blight. In 1970, blight destroyed an estimated 15 percent of the entire U.S. corn crop; in some places, the loss was 100 percent. The lesson was clear. In the words of one plant pathologist, "never again should a major cultivated species be molded into such uniformity that it is so universally vulnerable to attack by a pathogen, an insect, or environmental stress."[7]

By far the most well-known disaster attributable to genetic uniformity is the Irish potato famine of 1849. Potatoes came to Ireland in the sixteenth century, and by the nineteenth century, Irish peasants depended on them as their primary source of food and livelihood. As Catharina Japikse writes, "potatoes provided good nutrition, so diseases like scurvy and pellagra were uncommon. They were easy to grow, requiring a minimum of labor, training, and technology—a spade was the only tool needed. Storage was simple; the tubers were kept in pits in the ground and dug up as needed. Also, potatoes produce more calories per acre than any other crop that would grow in northern Europe. This was important to the Irish poor, who owned little, if any, of their own land. Often, a whole family could live for a year on just one acre's worth."[8] In the words of historian Cecil Woodham-Smith, potatoes were "the most universal of foods. Pigs, cattle and fowls could be reared on it, using the tubers which were too small for everyday use; it was simple to cook; it produced fine children; as a diet, it did not pall."[9] Potatoes enabled the population in Ireland to burgeon. But in 1845, a mold called *Phytophthora infestans* made its way to Ireland and, aided by moist weather conditions, attacked the potato crop. The 1845 harvest was substantially affected; in 1846 and 1848, the crop failure was total. Estimates vary, but as many as 1.5 million people are believed to have died of starvation and related diseases in the ensuing famine. A major reason why the blight was so devastating was that the tenant farmers, who only had small plots of land for their own use, had planted only one variety of potato, the "lumper," which, although not the best tasting, gave the highest yield per acre. When the blight struck, there were no resistant varieties.

Bananas are a food staple for an estimated half billion people in Asia and Africa. Wild bananas have large, hard seeds, but long ago, sterile, mutant versions were discovered that have small seeds, making them better for cooking and eating. Virtually all cultivated bananas are derived from these mutants, leading one science writer to describe the banana as "a freakish, doped-up, mutant clone which hasn't had sex for thousands of years."[10] Since these bananas are sterile, farmers propagate them by replanting their shoots. Until the 1950s, the dominant variety of banana was the Gros Michel, but it was killed off by a fungus called Panama disease. Instead of replacing it with a genetically diverse group of banana varieties, some of which might be resistant to future diseases, planters took the more profitable route and once again planted a single variety, the Cavendish. It now takes massive amounts of pesticides to keep a new fungal disease, called Black Sigatoka, at bay, and a new version of Panama disease has appeared that is not vulnerable to the pesticides that are now in use. Plant scientists are frantically searching for natural varieties of the banana that carry resistance to these diseases, but for the most part, what they have found has a different flavor and texture that may make them unacceptable to consumers. Ironically, the survival of the banana may lie in genetic engineering to make it resistant to fungal diseases, but this may significantly depress sales because of the widespread opposition to genetically modified foods.

Despite these historical lessons, crop diversity continues to decrease. Over the years, humans have relied on an estimated 7,000 plant species, yet only about 150 plant species are now cultivated, and most humans live off of no more than a dozen. Over 95 percent of the peas grown in the United States, for example, come from just nine varieties.[11]

The vulnerability of genetically homogenous crops raises the question of how serious a threat would a major loss of diversity in our own gene pool be to survival of the human species. In part, the answer can be derived from simple mathematics. If enough individuals were engineered with sufficiently similar genetic endowments, diversity would be lost. So how likely is it that parents would all make the same decisions about what modifications to install in their children? Chapter 4 laid out some reasons why this might occur, namely, the degree to which parents are influenced by cultural biases and by the behavior and preferences of social elites. If parents saw other parents installing the same genetic traits in their offspring, particularly in families that they envied or admired, they might feel pressure to do the same.

Another impetus for parents to make similar decisions would be the influence of the physicians and geneticists who performed the genetic engineering. The professional groups to which they belong constantly issue recommendations for how they should practice. For example, the American Society for Reproductive Medicine issues guidelines for what genetic testing donors should undergo in order to make sure their eggs and sperm are free from genetic disorders;[12] the American Academy of Pediatrics has policies on what conditions newborns should be screened for;[13] the American Society of Human Genetics, American College of Medical Genetics, and the American Medical Association (AMA) give rules for genetic testing for children and adolescents;[14] and the AMA tells its members how to perform gene therapy[15] and under what circumstances it would be unethical to use genetic engineering to enhance human traits.[16]

Conceivably, groups like this might come out with guidelines about what specific modifications should and should not be made, and physicians and geneticists would face strong pressure to conform to association policies. In 1994, when the first gene associated with a major risk for breast cancer was discovered and could be detected in a person's DNA, an NIH advisory committee declared that it was premature for clinicians to offer the test to their patients outside the context of a research study.[17] But the members of the advisory committee were genetic researchers, and their recommendation did not sit well with the American Society of Clinical Oncology (ASCO), whose members treat cancer patients. Accordingly, ASCO issued its own recommendation that the test "should be made available to selected patients as part of the preventive oncologic care of families."[18] Unlike the NIH pronouncement, the ASCO recommendation had teeth: not only might a failure to follow the recommendation be used as evidence of malpractice, but clinicians who failed to offer the test to their patients might be found to have violated the society's policies, which could lead to their being expelled from membership, which in turn could threaten their professional licenses and their positions on hospital and clinic staffs.

Marketing efforts by technology manufacturers and suppliers also will affect decisions about evolutionary engineering. The pharmaceutical industry spends more than $50 billion a year to promote its drug products,[19] and additional money is spent by other industries related to genetic engineering. Much of this effort is directed at health care professionals, but ever since the FDA relaxed its marketing rules in 1995, corporations have been aiming TV ads at consumers. Between 1996 and 2005, their expenditures on direct-to-consumer ads for prescription drugs alone increased by over 300 percent.[20] Added to this is the so-called technological imperative, the desire by health care professionals to employ the latest technological advances.[21] These factors may encourage genetics professionals to recommend similar choices to parents and motivate parents to request similar modifications for their offspring.

Another reason that many parents can be expected to make the same genetic choices is that they will want to provide their offspring with the most beneficial modifications, and there is bound to be a considerable degree of consensus on which ones were likely to lead to personal happiness and social success. Furthermore, there may not be many alternatives to choose from, since genetic engineering will be perfected gradually, and it will take time for a large number of safe and effective genetic changes to become available. Moreover, corporate developers can be expected to focus on the most lucrative genetic manipulations. At the outset, then, parents faced with a limited number of alternatives may opt for the same packages.

Parents also may face strong external pressures to select specific modifications. The state of the domestic economy may color their choices as they try to anticipate future growth sectors and equip their offspring with abilities that will translate into marketable skills. Employers and even entire industries hoping to create ideal future workforces may try to influence parental decision making through the media and by helping to pay for the modifications that they think will prove profitable, especially where there is a tradition of children pursuing the same careers as their parents. Parents are likely to be especially eager to secure favorable educational opportunities for their children. Elite educational institutions may declare what traits they are looking for in students, much as they now reveal the average GPAs and entrance examination scores of the students they accept, and parents might try to anticipate these educational preferences and install them in their children. You may recall the scandal involving investment analyst Jack Grubman. According to a complaint by the Securities and Exchange Commission, in 1999 Grubman wanted to enroll his children in the 92 Street Y, a prestigious Manhattan private school. When Sanford Weil, the CEO of Citigroup and a member of the board of directors of AT&T, asked Grubman to consider upgrading his assessment of AT&T's financial prospects, Grubman sought a quid pro quo. Weil wrote a recommendation letter for Grubman's children to the 92 Street Y, Grubman's children were admitted, he upgraded his assessment of AT&T, and Citigroup pledged a $1 million gift to the school.[22] (The Securities and Exchange Commission later fined Grubman and permanently barred him from the securities industry.) It doesn't seem far-fetched to imagine Grubman and others like him buying certain engineered talents for their children to improve their chances of being admitted to select schools and colleges.

Society may even come to regard certain genetic modifications as indicative of good parenting. Recall from chapter 4 Julian Savulescu's "Principle of Procreative Beneficence."[23] Parents could face social censure if they refused to select the most desirable embryos from the collection that had been fertilized in vitro, if they rejected an opportunity to genetically engineer their children, or if they installed odd or unpopular traits. In some communities, such as those with strong religious beliefs, the pressure to conform to group norms regarding genetics may be irresistible. As mentioned in the introduction, an example is the ultra-Orthodox Hassidic Jewish community of New York, where marriages are still arranged by matchmakers and require the permission of the chief rabbi. To stamp out Tay-Sachs, a recessive, fatal genetic disease that strikes infants, a program called Chevra Dor Yeshorim ("Association of an Upright Generation") screens adolescents for the Tay-Sachs mutation. Before a marriage can take place, the matchmaker calls the program with the identification numbers of the bride and groom, the program checks their test results, and if they both carry the gene for Tay-Sachs and therefore are at risk of conceiving a child with the disorder, the information is relayed back to the matchmaker and the rabbi and the marriage plans are scrapped. While Chevra Dor Yeshorim only involves genetic testing and mate selection, groups with similar authoritarian structures may embrace evolutionary engineering, and in those communities, communal attitudes may virtually dictate parental decisions. Later we will discuss the challenges created by one type of group, transhumanist cults.

A crucial question regarding evolutionary engineering is the role of government. The previous chapter mentioned the possibility that opponents of evolutionary engineering would attempt to ban it by law, but there are many reasons why the government instead someday might encourage or even require parents to modify their children in certain ways. The government might attempt to justify its directives as a public health measure. The heads of several science and policy programs at the AAAS, for example, have declared that we owe it to our descendants to make genetic changes that are predicted to improve their lives, such as eradicating disease genes.[24]

Another reason the government might encourage genetic engineering is global competition. James Cameron, who runs the Institute for Global Futures, is convinced that enhancing human traits will be absolutely essential in order to enable the United States to compete successfully in the global economy, particularly against countries that employ genetic engineering to enhance their workforces.

One country that springs to mind is China. The Chinese government has quite a track record of controlling its citizens' reproductive decision making based on genetic factors. In 1994, it passed a law requiring every couple to undergo a premarital medical examination. If the exam reveals a "genetic disease of a serious nature which is considered to be inappropriate for child bearing from a medical point of view," the couple may marry only if they agree to take measures to avoid having children, such as use of long-term contraceptives or sterilization. Moreover, termination of pregnancy "must be advised" when prenatal genetic testing shows that a fetus has a "defect of a serious nature" or a "genetic disease of a serious nature."[25] China even has intentionally bred gifted athletes. As correspondent Brook Larmer writes in his book *Operation Yao Ming*, "when Chinese athletes reach the end of their playing days, they are never truly released from their obligation to the state. Until recently the sports system automatically absorbed most retired athletes as coaches or administrators, who passed on their knowledge to the next generation. If they happened to be extraordinarily tall or talented, they were expected to pass along something even more fundamental: their genes."[26] Hence, when tall female basketball player Da Fang was allowed to retire, a marriage was arranged between her and another player, Yao Zhiyuan, who was 6 feet 8 inches tall, and as planned, they gave birth to a giant, who was named Yao Ming. When Yao Ming was eight, Chinese sports officials enrolled him in a basketball training academy. Eventually, he and his parents moved to the United States, where he now plays for the Houston Rockets. The type of evolutionary engineering that the Chinese employed in producing Yao Ming was old-fashioned selective breeding, but his story is a harbinger of how China might view more advanced forms of genetic engineering as they become available.

Another area of government interest in genetic engineering is likely to be in connection with the military. In 2001, the Committee on Opportunities in Biotechnology for Future Army Applications of the Board on Army Science and Technology at the National Research Council issued a report calling for the U.S. Army to "lead the way in laying ground-work for the open, disciplined use of genomic data to enhance soldiers' health and improve their performance on the battlefield."[27] In 2002, a report by the Department of Defense Information Assurance and Analysis Center observed that "because genomics [sic] information offers clues to improving human performance it could provide the Army with means of increasing combat effectiveness."[28] Most recently, in December 2010, the JASONs, a group of scientific advisors to the military, issued a report entitled "The $100 Genome: Implications for the DoD" that outlined an ambitious plan to employ genomic technologies to "enhance medical status and improve treatment outcomes," enhance "health, readiness, and performance of military personnel," and "know the genetic identities of an adversary."[29] The U.S. military is reported to now be spending $100 million a year on research designed to counteract sleep deprivation alone,[30] along with many other programs aimed at enhancing soldier performance. So far as is known, the U.S. government is not yet employing genetic engineering, but if the technology improves, that day may not be far off. It also is difficult to imagine that the government's interest in enhancing performance would lead it to engage in germ line genetic engineering, for example, attempting to create children who would make ideal warriors as adults. In families with long traditions of military service, however, parents might be interested in installing traits that were conducive to a successful military career, and it is not that far-fetched to think that the government might encourage this practice, perhaps even subsidizing it.

Whether the government should play a role in parental decisions about what sorts of children to have, and if so, what that role should be, turns out to be one of most highly charged issues raised by evolutionary engineering, for the government intervened in this realm once before, and the result was one of the more horrific chapters of American, and ultimately world, history: the eugenics movement. The term "eugenics" was coined in 1883 by an Englishman named Frances Galton, who defined it as "the science which deals with all influences that improve the inborn qualities of a race; also with those that develop them to the utmost advantage."[31] The movement Galton started found its way across the Atlantic, where it received substantial financial support from leading citizens, including the Harriman, Carnegie, and Rockefeller families. By 1931, after the U.S. Supreme Court had endorsed eugenic sterilization in the case *Buck v. Bell*, twenty-eight states had enacted laws permitting criminals and the mentally unfit to be involuntary sterilized so that their "inferior" genes would not be passed on to future generations. Approximately 3,000 of these operations were reported to have been performed each year prior to World War II, and many more took place that were not reported.

The American eugenics movement exerted a major influence on Adolph Hitler, who heard about it in jail in 1924 when he was writing *Mein Kampf,* and the first law passed after he came to power in 1933 was a sterilization provision modeled on the Virginia statute that the U.S. Supreme Court had upheld in the *Buck* case. Hitler's original idea was to sterilize inferior populations to make them gradually disappear, but when the Germans ran out of hospital beds for wounded soldiers at the start of World War II, Hitler ordered mental patients to be killed by lethal injection in order to free up their beds. Ultimately, the Nazi eugenics program became the Holocaust.

In addition to promoting involuntary sterilization to curb the spread of "bad" genes, the eugenics movement also endeavored to encourage the birth of children with "good" genes. Church sermons promoting marriages between genetically superior persons were given awards by the American Eugenics Society. Families in the 1920s toured "Fitter Families Exhibits" at state fairs and entered themselves in "Perfect Baby Competitions." Students with good genes won "Goodly Heritage" medals. The

Nazis embraced positive eugenics as well, embodied in their "Lebensborn" program, in which Aryan women were mated with SS soldiers and their offspring raised in special foster families.

The Holocaust naturally dampened enthusiasm for eugenics, but the movement was not eliminated altogether. States continued to sterilize people without their consent: 186 women in North Carolina between 1948 and 1955; 574 in Georgia, North Carolina, and Virginia in 1954; and 104 inmates, 102 of whom were African Americans, at a single South Carolina institution between 1949 and 1960. Nor were the United States and Germany the only countries with eugenics programs. Sweden sterilized 60,000 people between 1935 and 1976. The victims, mainly women, were labeled as genetically or racially "inferior" and were typically poor, learning disabled, or non-Nordic. One of the major justifications for the Swedish sterilization program was that it would reduce the cost of the Swedish welfare state by reducing the number of people who would have to be supported. One woman is reported to have been sterilized because she could not master her confirmation studies well enough to satisfy her priest. Another who was sterilized after she was judged to be mentally slow as a child because she could not read the blackboard later turned out to merely need eyeglasses.[32] In 1999, the Swedish government agreed to pay $20,780 to each surviving person who had been sterilized.

Eugenics is now said to be a thing of the past. "Once favored by both the left and the right," transhumanist Nick Bostrom assures us, "the last century's government-sponsored coercive eugenics programs . . . have been thoroughly discredited."[33] According to the authors of the philosophical text *From Chance to Choice: Genetics and Justice*, the eugenics movement "is largely remembered for its shoddy science, the blatant race and class biases of many of its leading advocates, and its cruel program of segregation and, later, sterilization of hundreds of thousands of vulnerable people who were judged to have substandard genes."[34] It therefore might seem unthinkable that the U.S. government would embark on another eugenics effort in which it sought to dictate parental decisions about how to genetically engineer their children. "Because people are likely to differ profoundly in their attitudes towards human enhancement technologies," says Bostrom, "it is crucial that no one solution be imposed on everyone from above but that individuals get to consult their own consciences as to what is right for themselves and their families."[35]

But is eugenics really dead? In a book chapter I recently wrote entitled "Modern Eugenics and the Law," I described a large number of government programs that are consistent with eugenic objectives.[36] Several government initiatives, for example, discourage "undesirables" from having children. California, Georgia, Florida, and Montana have laws that mandate castration for repeat male child molesters; although the ostensible purpose of these enactments is to prevent child molestation by reducing the offenders' sex drive, it also prevents them from reproducing. Stopping poor people from having children was one of the main objectives of twentieth-century eugenicists, prompting Margaret Sanger to found Planned Parenthood; today, the government spends more than $1 billion a year on family planning programs that provide contraceptives and sterilizations to the poor. Another government policy that discourages the poor from having children are the "family caps" in 24 state welfare programs, which halt increases in welfare payments to poor families once they exceed a certain size. On the "positive eugenics" side, child tax credits encourage better-off families to have more children, since the poor don't pay income taxes, and the more children the taxpayer family has, the larger the number of credits that it gets to claim. State fairs also have recently resurfaced in connection with genetics; at the Minnesota State Fair in the summer of 2010, researchers from the University of Minnesota collected DNA from children for research purposes.[37]

The government program that most closely resembles eugenics, however, is newborn screening. The program began in the 1960s after physician Robert Guthrie developed a test for PKU, a metabolic disorder that can be managed effectively if it is detected immediately after birth and the child's diet avoids foods containing the amino acid phenylalanine, including meat, chicken, fish, eggs, nuts, and dairy products. As soon as he developed the PKU test, Guthrie began urging state public health officials to use it to test newborns. By 1973, newborn screening was compulsory in 43 states; now it is compulsory in all. Most of the conditions tested for, like PKU, affect children and are treatable, and the screening programs are justified on the basis that they identify children who will benefit from early treatment.

But the development of faster and cheaper testing methods has led to calls for expanding the screening programs to identify diseases for which there are no highly effective remedies, as well as those that only manifest symptoms in adulthood. Proponents argue that this could spare families years of uncertainty once symptoms emerged, alert them to be on the watch for new medical discoveries to treat affected children, increase children's access to medical care that might alleviate, if not cure, their conditions, and facilitate enrolling children in studies researching their disorders. None of these rationales raise eugenics issues, but unlimited genetic screening also is defended on the basis that it can provide "knowledge on which to base reproductive decision-making years before a disease would be diagnosed for the affected child."[38] In short, children should be genetically tested so that their parents can avoid having another child with a genetic disorder, and this should be *mandated* by the government. (In 2005, the Nebraska Supreme Court ruled against parents who tried to block public health officials from testing their newborn because it offended the parents' religious beliefs. In ordering the child to be screened over the parents' religious objections, the court cited the need to address "the potential social burdens created by children who are not identified and treated.") A mandatory government program aimed at preventing the births of children with disabilities doesn't simply bear a resemblance to eugenics; it *is* eugenics.

In addition, if the government can require genetic screening in order to discourage parents from passing defective genes on to their offspring, it may go further and encourage parents to eradicate those genes by modifying their children's DNA directly. This could have a significant impact on the genetic makeup of future generations, including causing a loss of genetic diversity if too many children's genes were engineered in the same way.

But the government is in a position to affect parental decisions about genetic engineering in others ways besides mandating newborn screening and offering tax advantages to parents who follow the government's recommended modification regime. The government also maintains significant control over the introduction of new medical technologies. With very few exceptions, the FDA must approve a new drug, medical device, or "biologic" before it is marketed, and the agency has used this authority to regulate the use of gene therapy. To obtain FDA approval, the manufacturer or "sponsor" must demonstrate that the technology is safe and efficacious, and human experiments may not take place without the agency's permission. Therefore, the options legally available to parents will be limited to those that the FDA has found to meet its statutory standards.

The weasel word, of course, is "legally." Assuming that the FDA now has or in the future is given jurisdiction over human genetic engineering, a genetic modification that was not approved by the FDA would be illegal. But as noted earlier, parents might attempt to get around that by obtaining the technology on the domestic black market or in other countries. Moreover, once the FDA approves something for one purpose, it is legal for a doctor to give it to patients for another purpose, a practice called "off-label use." Since the basic methods for engineering health-promoting changes in children's DNA such as IVF and tinkering with embryonic DNA are likely to be the same ones that would be used to genetically enhance them, physicians' off-label authority hampers the government's ability to prevent genetic engineering from being used for nonmedical purposes. Suppose that, as some critics of evolutionary engineering have recommended, parents were left free to use genetic engineering to eliminate harmful genes from their children's DNA, but they were not allowed to use the technology to enhance their children, that is, to install traits or abilities that made children better than normal, at least not unless it was to improve their health (to rule out things like vaccinations, which give people a better than normal ability to fight off disease). If the FDA approved a genetic technique for health purposes, under the law as it now stands, the FDA would not be able to prevent licensed physicians from using the technique for enhancement purposes. The only way to stop it from being employed for enhancement would be to make it illegal altogether, including using it to treat disease, and such a draconian step is likely to be politically unacceptable, especially if the technique was highly effective and the disease in question was serious or life threatening.

Yet while black markets, genetic tourism, and off-label use may complicate its efforts, one way or another the government is bound to try to restrict parental freedom to design their children to their own taste. Some types of genetic engineering may not meet FDA safety and efficacy standards, especially when they are first developed. Parents may adhere to the FDA's rules because they are law abiding, or because they are worried about the safety record of black marketers or foreign providers. Much of the risk associated with the use of performance-enhancing drugs by athletes, for example, may be due to their need to rely on illicit sources. If it felt it necessary, Congress could make it unlawful for doctors to use genetic engineering for off-label purposes. Finally, opponents of evolutionary engineering may succeed in making some types off-limits, such as mixing animal and human DNA.

In terms of the effect of evolutionary engineering on the genetic diversity of the species, the question is how much all of these factors—

pressures to conform, similarity of parental objectives, and government restrictions—would shape the human gene pool. Clearly, the more alike parents acted in altering their offspring, the less diverse the gene pool would be. But biologist Jon Gordon thinks that the natural process of random genetic mutation will continue to produce sufficient genetic diversity to protect the species against sudden environmental changes. Any loss of diversity resulting from intentional engineering, he argues, "would be swamped by the random attempts of Mother Nature."[39] The late political philosopher Robert Nozick even called for a "genetic supermarket" in which parents who wished to engineer their offspring could make individual selections similar to buying groceries. "This supermarket system," he said, "has the great virtue that it involves no centralized decision fixing the future human type(s)."[40] There are additional reasons to be optimistic that a sufficient amount of diversity can be maintained. Although the exact figure is not yet known, it is believed that there is considerable genetic variation between individuals, perhaps as much as between humans and their nearest primate relatives, chimpanzees.[41] It would take a great deal of copycat genetic engineering to significantly reduce this variation. Moreover, human populations differ widely in their cultural attitudes and resources, so that if some groups became genetically homogeneous, others might not, either because they didn't want to or because they couldn't afford to, and evolutionary biologists point out that, so long as the engineered and natural populations interbreed, a substantial amount of diversity will persist.

Despite these reasons for optimism, however, there are still reasons to fear that too much genetic diversity will be lost. Philosopher Jonathan Glover warns that, by letting parents select the characteristics of their children, Nozick's genetic supermarket could end up reducing rather than increasing genetic diversity. "The influence of fashion or of shared values," Glover warns, "might make for a small number of types on which choices would converge."[42] Furthermore, new genetic tests are being marketed every day, and they are costing less and less. In June 2010, one company announced that it would decode a person's entire genome for $14,500;[43] the NIH is committed to lowering the cost to $1,000 as soon as possible.[44] A recent report on the future use of genetic technologies by the military anticipates "the $100 genome."[45] As the price drops, more parents will be able to afford genetic testing for embryos and fetuses, and fewer children will be born with genetic differences that cause genetic disorders. As genetic engineering is perfected, parents will have additional ways to produce offspring with similar sets of optimal genes. Even if only a few portions of DNA were altered by many parents in the same way, these may turn out to be regions that code for traits that significantly increase the survival and reproductive success of their off-spring. Moreover, we saw earlier that one gene may perform many functions, called pleiotropy. If the areas of DNA that parents altered were highly pleiotropic, this could have an unexpectedly large effect on how well suited their children were to their environments. Evolutionary biologists worry, in fact, that even a small number of changes in pleiotropic genes could have dramatic effects on the human species.[46]

The more that the human species resembles a genetic "monoculture," the less it would be able to adapt to changes in the environment. This may not be a serious problem if environmental changes take place gradually enough, since the same tools of genetic engineering that reduced genetic diversity might be able to restore it, or at least to equip people with the adaptations they needed to survive. If too many people proved susceptible to a new pathogen, for example, resistance to that pathogen could be installed genetically. Researchers already are constructing biobanks that will store large amounts of human, animal, and plant DNA samples, such as the "Frozen Ark" project in Great Britain, which focuses on storing genetic material from animals threatened with extinction.[47] Even if living organisms lost their genetic diversity, it could be restored using DNA from these repositories. The prospect of harm from genetic engineering therefore doesn't worry Jonathan Glover too much. It's not as if we're making irrevocable choices for all time, he points out.

But if the new environmental challenges occurred suddenly and dramatically, there might not be time for enough individuals to adapt even at the accelerated pace made possible by genetic engineering. Lengthy experimentation may be needed to identify the genetic modifications that are required for survival and to learn how to make them successfully. Bear in mind that, as former UCLA geophysics professor Didier Sornette points out, the term "environment change" covers a broad range of potential cataclysms, ranging from "large natural catastrophes, such as volcanic eruptions, hurricanes and tornadoes, landslides, avalanches, lightning strikes, catastrophic events of environmental degradation, to the failure of engineering structures, social unrest leading to large-scale strikes and upheaval, economic drawdowns on national and global scales, regional power blackouts, traffic gridlock, diseases and epidemics, etc."[48] At present, it is virtually impossible to predict with any accuracy when these types of catastrophes will occur. According to the U.S. Geological Survey, for example, "currently no organization or government or scientist is capable of successfully predicting the time and occurrence of an earthquake."[49] When asked if the field of predicting volcanic eruptions was "well on its way to becoming an exact science" or "still in its infancy," one volcanologist replied, "Well, when it works, it's well on its way. When we have a spectacular failure, it's in its infancy."[50] If economic forecasting is so accurate, how is it that the economy lost $500 billion in real gross domestic product between 2008 and 2009?[51] Without enough of an advanced warning, even sophisticated techniques for genetic engineering may be unable to save the human lineage.

Finally, even a genetically diverse population can be threatened by cumulative, deleterious changes in its collective DNA. One pessimist, for example, is biologist John B. Fagan. He warns that "the inevitable slip of the genetic scalpel by even the best-intentioned scientist will bring harm not only to one person, but to all subsequent offspring," and adds that "scientists acknowledge that these mistakes are unavoidable. Therefore, if applied widely, germline engineering will progressively corrupt the blueprint of our species with genetic errors. These will irreversibly burden future generations with new genetic diseases, causing millions to suffer. Such manufacturing defects cannot be recalled."[52] Eventually a tipping point could be reached at which the species embodying the lineage became so weakened and disease ridden that it was incapable of survival.

In addition to a loss of genetic diversity and cumulative, inborn genetic errors that make a lineage vulnerable to a sudden environmental insult, other conditions are potentially lethal to it. One is extremely large body size. Large animals require more resources to sustain their metabolisms, so they are more vulnerable to breaks in their food chain. They also tend to reproduce less often.[53] Since being a taller human confers social benefits, chapter 4 raised the possibility that parents might genetically engineer taller offspring, but it is unlikely that they would want to create giants. If for some reason they did, and if the practice was common, then the resulting humans might become more vulnerable to extinction.

A more likely threat to the lineage is reproductive failure, the failure to reproduce in such a way that the lineage can perpetuate itself. This could happen for several reasons. Humans could slow their rate of reproduction voluntarily, bearing fewer children during their lifetimes, perhaps in response to shrinking economies. Europe is already seeing declines in population. At the current rate, by the mid-twenty-first century the Ukraine will have lost half its population, and birthrates are similarly shrinking in the Czech Republic, Italy, Poland, Russia, China, and parts of Germany, as well as in much of the Arab world.[54] (At the same time, however, birthrates are increasing dramatically in Sub-Saharan Africa.) In evolutionary jargon, a low birthrate is called a "slow life history," and lineages with a slower life history are at a greater risk of extinction because there are fewer opportunities for genetic change to produce individuals better adapted to variations in the changing environments. Some studies in fact show that a slow life history is a better predictor than size for which species became extinct during the large-scale die-off in the Late Pleistocene epoch, when "megafauna" such as mastodons, cave bears, and saber-toothed tigers disappeared.[55] In addition to deliberate actions to reduce family size, humans might become less fertile because of inadvertent changes produced by evolutionary engineering. Nicholas Wade, for example, speculates that birthrates may decline because parents opt for children with lighter skin color in sunny climates such as Africa, where the additional ultraviolet radiation would decrease fertility by reducing the body's production of folic acid.[56] (Wade claims that lighter skin makes evolutionary sense in northern places since it enables more vitamin D to be obtained from scarcer sunlight.) Evolutionary engineering also could inadvertently disrupt the complex genetic interactions that are required for reproduction.

Another danger to the survival of the lineage is gender imbalance. A significant lack of males or females could slow the birthrate and decrease genetic diversity by reducing the number of reproductive cells available for fertilization from the underrepresented sex. Moreover, competition for a scarce number of mates may have adverse health effects. A recent study of almost 8 million men over 50 years found that large imbalances decrease the life span of the overrepresented gender by 3 months at age 65. This may not seem like much, but it is comparable to the *increase* in life span that older people enjoy if they make all of the behavioral changes that are commonly recommended for increasing longevity.[57]

Gender imbalance already is widespread in Asia, where ultrasound and other techniques have allowed parents to identify the sex of their fetuses and abort the gender that is undesirable. The aborted fetuses are invariably female. In rural China, for example, sons are preferred because they are believed to make better farm workers, provide financial support to elderly parents, preserve the family name, and, in the past at least, bear the responsibility for the worship of ancestors.[58] This bias has ancient roots; a 3,000-year-old Chinese poem reads,

> When a son is born,
> Let him sleep on the bed,
> Clothe him with fine clothes,
> And give him jade to play with;
> When a daughter is born,
> Let her sleep on the ground,
> Wrap her in common wrappings,
> And give broken tiles for playthings.[59]

The Chinese government calculates that 118 boys are being born for every 100 girls. Journalist and author Martin Walker says that millions of Chinese males, whom the Chinese call "bare branches," may never find a mate,[60] and he cites a study stating that almost 90 million females may be "missing" in Asia as a consequence.[61] Indians prefer male children for the additional reasons that sons protect the family while daughters must be protected, and that sons add to the family's wealth while daughters decrease it because the family must provide their dowries.[62] As one Indian woman is reported to have said, "Better to pay $38 for an abortion now than $3800 for a dowry later on."[63] The United Arab Emirates has the greatest imbalance: 205 males for every 100 females[64] (due, no doubt, to the influx of male foreign workers). In the United States, a survey found that half of fertility clinics that test embryos before implantation reported that they permit parents to select the sex of their child.[65] Law professor Lori Andrews calls this "gynecide."[66]

So if potential harm to children and societal disruption aren't enough to make us want to halt the march toward evolutionary engineering, its potential to bring about the demise of the human lineage certainly seems like it should be sufficient. As the saying goes, and as opponents of evolutionary engineering might point out, don't fool with Mother Nature. The best way to ensure the survival of our species, they might say, is to let evolution proceed the old-fashioned way, at its own pace.

There is one problem with that argument, however: does natural evolution actually work so well? As the great evolutionary biologist Ernst Mayr points out, if natural evolution is so good at enabling species to adapt to their environments, how come 99.9 percent or more of all evolutionary lines are extinct? Mayr gives a host of reasons why natural evolution does such a poor job of assuring evolutionary success: gene pools contain only so much genetic variation; individual organisms have a limited amount of genetic and nongenetic flexibility with which to respond to changes in the environment; natural selection doesn't operate once an organism is too old to reproduce; evolution does not eliminate all physical deficiencies, such as those that are due to "phyletic" constraints, which are the product of the genetics that govern early development (Mayr gives the example of the proximity in humans of the trachea and the esophagus, which permits choking on foods and liquids); and finally, one must account for chance events, such as environmental surprises and the random exchange of genetic material during reproductive cell division.[67]

The precariousness of group survival under natural evolutionary conditions, moreover, is a major motivation for transhumanism. "We should not engage in biology worship," declares Julian Savulescu. "Our biology is not sacrosanct. We should change it to make our lives longer and better."[68] James Watson, who along with colleague Francis Crick discovered the structure of DNA in 1953, puts it more acerbically: "I just can't indicate how silly I think [the sanctity of the human gene pool] is. I mean, sure, we have great respect for the human species. We like each other. We'd like to be better, and we take great pleasure in great achievements by other people. But evolution can be just damn cruel, and to say that we've got a perfect genome and there's some sanctity to it . . . [is] utter silliness."[69]

So perhaps the real question is not whether evolutionary engineering poses a threat to the preservation of the human lineage, but whether it poses a greater threat to the lineage than natural evolution. But this question assumes that we have a good grasp of the workings of natural evolution. As the foregoing discussion suggests, we have some notions about what sorts of conditions create threats to the survival of a species or lineage. But how well do we really understand natural evolution?

CHAPTER SEVEN

Evolution by Nature or by
Human Design?

Leon Kass, an outspoken opponent of genetic engineering, once summed up his aversion to evolutionary engineering with the following observation: "Though well-equipped, we know not who we are or where we are going."[1] Kass's aphorisms often sacrifice clarity for grandiloquence, but it is usually worth the effort to figure out what he is getting at. What are we ignorant of? Well, for starters, despite 150 years since Darwin, there is still a lot we don't know about the process of evolution.

Our ignorance of the details does not mean that the basic concept of evolution is in doubt. It is not a confirmation of the beliefs of the creationists, although the lines between religion and science are blurred by the many evolutionary scientists—from the Jesuit paleontologist Teilhard de Chardin to Frances Collins, the current head of the NIH—who believe that evolution is one of God's instruments. Regardless of their belief systems, however, scientists have not been able to fill in many of the chapters of the human evolutionary record, and where they have tried, the pages are almost illegible as a result of cross-outs and erasures, passionate and often rude scrawls of rejection and rebuttal, and wording blurred by tears of frustration. This has profound implications for evolutionary engineering: If we do not have a good understanding of our evolutionary past, how can we predict our evolutionary future? If we cannot clearly mark the evolutionary path that brought us to what we are today, how can we avoid straying into evolutionary peril tomorrow? And without greater knowledge of the process of natural evolution, how can any side in the debate—transhumanists, evolutionary biologists, conservative skeptics, or creationists—say with any confidence that evolutionary engineering poses a greater threat to the preservation of the human lineage than allowing a more natural approach to evolution take its course?

The uncertainties about our evolutionary past may come as a surprise to you. Walk through any museum of natural history and everything seems wonderfully straightforward: A mixture of chemicals and solar energy produced life in the seas, some fish eventually crawled up on land, dinosaurs gave way to mammals, humans gained the brains and technologies to rule the roost, and all this operated according to settled and well-understood principles and processes. But it turns out that we know far less about the clearly labeled bones and artifacts in those museum cases, the parade of progressively advanced skeletons, the dioramas of prehistoric animals, and "early human life on the savannah" than you might think.

There is even debate about the basic operation of evolution. Scientists generally agree that it involves four processes: random mutation, natural selection, gene flow, and genetic drift. Random mutation refers to radiation and other external insults, as well as cell division, that cause mutations in DNA. By occasionally producing differences in certain organisms that make them more likely than others to reproduce successfully under changing environmental conditions, random mutation gives rise to the second process, which Darwin called "natural selection." Organisms also sometimes reproduce more successfully owing to DNA that they acquire by interbreeding with other populations, a process called "gene flow." Finally, there is "genetic drift," where genetic differences from one generation to the next occur randomly during the act of reproduction.

But once past this basic framework, the uncertainty begins. Which of these processes is most important? Darwin and the naturalists who followed his lead emphasized natural selection. To them, virtually every change in an organism was subject to its operation. In Darwin's case, this is understandable, since he had little awareness of the tiny packages of inherited information called genes and therefore would have had trouble grasping that they could change over time as a result of the operation of other forces. (There is a disagreement about whether Darwin knew about the work of Gregor Mendel, the Austrian monk who was Darwin's contemporary and who first perceived the operation of the process of genetic inheritance in pea plants.)

By the early twentieth century, researchers had discovered pairs of structures within reproductive cells that separated and split when the cells divided, calling them "chromosomes"—a term meaning "colored things"—because the researchers stained them with dyes in order to make them visible under a microscope. Other scientists soon figured out that the units responsible for Mendel's patterns of inheritance were elements of the chromosomes, and they gave them the name "genes," a term a Danish botanist had derived from "pangenesis," which Darwin had used to describe the mechanism of heredity. By the late 1940s, Darwin's and Mendel's insights had been woven together into an amalgamation called "the modern synthesis," a term originated by Julian Huxley, who, it will be recalled, also came up with "transhumanism."

But the modern synthesis didn't settle the question of whether natural selection was the predominant factor producing evolutionary change. Some evolutionary scientists argued that it couldn't be because most random genetic mutations, the grist for natural selection's evolutionary mill, have no positive effect on an organism's reproductive success. In their opinion, the primary evolutionary process had to be genetic drift, the random process of genetic change over time. By the late 1980s, the drift argument seemed to be winning out, despite the inconvenient problem that it could not actually be proven, since, as the authors of a classic biology text observe, "to do this it would be necessary to show that selection has definitely NOT operated, which is impossible."[2] More recently, however, the defenders of natural selection have rallied, and, as in Darwin's time, they appear once again to enjoy the upper hand.[3]

The disagreement about natural selection versus genetic drift spills over into a number of other disputes. What is the fundamental unit of evolution? Darwin thought it was the individual organism, which natural selection makes more or less reproductively successful. A British zoologist with the unusual name of Vero Copner Wynne-Edwards argued that the basic evolutionary unit was groups of organisms. The American biologist George Williams insisted that it was the gene, and he gained a powerful ally in Richard Dawkins, author of the book *The Selfish Gene*.[4] Many scientists now accept that there is no single unit of evolution, and that its processes operate on genes and groups as well as on organisms, but others are not convinced.

One of the brickbats that the gene-versus-organism-versus-group combatants hurl at one another is their competing views on the abruptness of evolution. The original evolutionary researchers based their theories on the available physical evidence, which were fossils. The fossil record shows plenty of evolutionary change, but it appears to take place suddenly, even explosively: here's a skeleton of a fish with fins, there's one with legs. But this doesn't make sense in terms of how the forces of evolution are supposed to operate. Random mutation, natural selection, gene flow, and genetic drift ought to make the pace of evolution gradual, even glacial, rather than explosive. As Ernst Mayr observes, this slow pace would be expected because, absent a dramatic change in the environment, the existing forms of life would have become highly adapted; therefore, there would be no reason for evolutionary leaps. "Owing to the hundreds or thousands of generations that have undergone preceding selection," explains Mayr, "a natural population will be close to the optimal genotype. . . . [U]nless there has been a major change in the environment, the optimal phenotype is most likely that of the immediately preceding generations. All the mutations of which this genotype is capable and that could lead to an improvement of this standard phenotype have already been incorporated in previous generations."[5] But then what about those fossils? Well, maybe the fossils simply can't capture change that slow. But what prevents them from doing so?

The mystery deepens with our growing knowledge of human genetics. Abrupt evolutionary shifts could occur if DNA mutated quickly enough to produce monumental changes in an organism. Molecular geneticists can now measure how fast DNA mutates by examining the physical evidence, which once again are the fossils, some of which contain sufficient amounts of ancient but still readable DNA to be analyzed. The geneticists' calculations show that the rate of random change in DNA over time is far too slow to produce the explosive bursts of evolutionary change reflected in the fossil record.

But if both the theory of evolution and the DNA evidence show that evolutionary change is slow, why isn't this evident in the fossils themselves, rather than just in their DNA? Enter the creationists. This is not an enigma, they explain, but a miracle, evidence, if not conclusive proof, that God exists.

Population geneticists disagree with the foregoing interpretation of fossil DNA. You can't just look at the physical DNA evidence, they argue. You have to understand how evolution occurs in large, interbreeding populations. At this level, evolution takes place not so much abruptly as *sporadically*. In other words, big changes in large groups are very rare, and the rarer something is, the less likely we would expect to find evidence of it in the fossil record. Moreover, when changes that significantly improve reproductive fitness do take place, they start in a small way but spread quickly throughout the population. Therefore, there won't be a lot of transitional organisms to be deposited as fossils. This sounds plausible, but if the affected population is large enough, shouldn't we find at least some transitional fossils? Not necessarily, reply the population geneticists. The extent of fossil excavation to date is actually quite limited, fossils can't capture many types of distinguishing characteristics, and the geologic conditions necessary for fossils to be preserved over long periods of time are not that common.

Even well-settled understandings can unravel very quickly. Long before the existence of genes was known, the French naturalist Jean-Baptiste Lamarck, who lived in the late eighteenth and early nineteenth centuries, posited that environmental conditions could produce changes in organisms that were passed on to their offspring. For example, giraffes stretched their necks to reach higher vegetation, and so their calves were born with longer and longer necks. Lamarck was ridiculed by geneticists, who insisted that, although environmental conditions could affect which genes were inherited, they could not change the genes themselves. Giraffes stretching their necks did not affect the neck length of their offspring; instead, giraffes with longer necks could reach higher vegetation, which made them reproductively more successful than giraffes with shorter necks, and so more calves tended to be born with longer necks. Interestingly, however, the discovery of genetic switches and epigenetics, described in chapter 3, suggests that Lamarck was partly correct: perhaps genes cannot be altered directly by the environment, but behaviors triggered by environmental conditions do seem to be able to affect genetic switches in ways that appear to be inheritable.

Another fundamental difference is over how to classify living things. Until fairly recently, the prevailing approach was to identify species based on differences in the appearance of their fossils and to portray evolution as relatively linear. This approach employs a set of classifications ordered from the less to the more specific (e.g., kingdom or domain, phylum, class, order, family, genus, species) and conceives of human evolution as a continuous sequence from earlier, less humanlike organisms to modern *Homo sapiens*. A more modern, though still controversial, alternative, known as cladistics, uses classifications based on evolutionary ancestry. There are a number of different versions of the cladistic approach, but essentially it traces evolutionary pathways from common ancestors onward. In contrast to the earlier depiction of human evolution as a fairly straight line with a few branches representing nonhuman dead ends, the cladistic approach appears almost "bushy," with modern humans on the end of one of many evolutionary branches.

The differences between these approaches may sound highly technical and esoteric, but they have had important consequences for the study of evolution. For example, the linear approach had a major effect on how naturalists interpreted the humanlike fossils they found. Since the greatest achievement would be to find a new link in the human evolutionary chain—if possible, the fabled "missing link" between chimpanzees and humans—rather than the remains of just another nonhuman primate, naturalists tended to emphasize the humanlike features of their finds and minimize any apelike attributes.[6] This led to all sorts of confusion and misattribution. Furthermore, fossils that seemed to contradict the prevailing understanding of the order of human evolutionary development were relegated to nonhuman status. The classic example is the Taung child, Raymond Dart's 1925 discovery in South Africa of the fossilized skull of an infant that had a small brain and humanlike teeth. Since the dominant wisdom was that an enlarged human brain had preceded human teeth, Dart's claim that he had found an early member of the human family was widely rejected. Ironically, one of his chief detractors was Sir Arthur Keith, a Scottish anthropologist and naturalist who was connected with, and some say one of the perpetrators of, the great Piltdown man hoax, in which an orangutan jawbone was attached to a human skull and passed off as the missing link in 1912.

More important, the linear approach to classification reflected the conviction that evolution was progressive. "Evolution," observed bioethicist and former priest Joseph Fletcher in 1988, "is about greater sophistication, complexity, and functionality."[7] This understanding served a number of important purposes: it suited those who needed to see a bright line separating humans from other living things, especially from the disturbingly similar apes and monkeys; it promoted the Old Testament teaching that God gave Man dominion over the earth; and it comforted those who, despite accepting evolution, still believed in God, or at least that the universe was home to a powerful benign force.

The progressive theory was best captured by a foldout illustration in a 1965 Time-Life book depicting human evolution as a row of figures starting with a small, gibbon-like creature called *Pliopithecus* on the left and moving to the right through a succession of increasingly more upright and humanlike forms until it reaches *Homo sapiens*, striding confidently forward clutching a staff. The illustration is so iconic that it has its own name, "The March of Progress."

But the march of evolutionary progress turns out to be wishful thinking. Natural evolution is not progressive for the simple reason that it can't be. As Ernst Mayr said, "there is no known genetic mechanism that could produce goal-directed evolutionary processes."[8] Not only is there no such natural force, but there would be no point to it. "Evolution," says Nicholas Wade, "has no goal toward which progress might be made."[9]

This truth is well supported by the scientific evidence. In terms of adaptation, creatures much simpler than man are far more successful. "Fish still dominate the oceans," points out Ernst Mayr, and in most environments, with the exception of humans, rodents are more successful than primates. In fact, Mayr adds, "the earliest . . . organisms, the bacteria, are just about the most successful of all organisms, with a total biomass that may well exceed that of all other organisms combined."[10] To understand just how successful bacteria are as a form of life, bear in mind that, whereas biologically modern humans have only been around for 1.8 million years, bacteria have flourished for 3 billion years. Furthermore, far from acting like a conveyor belt for positive change, the history of evolution for the most part is one of stagnation: "Three billion years of unicellularity, followed by five million years of intense creativity and then capped by more than 500 million years of variation on set anatomical themes," observes Stephen J. Gould, "can scarcely be read as a predictable, inexorable or continuous trend towards progress or increasing complexity."[11] As for the primacy of humanity, says paleoanthropologist Ian Tattersall, "it is already evident that our species, far from being the pinnacle of the hominid evolutionary tree, is simply one more of its many terminal twigs."[12]

Understanding why people thought otherwise, however, holds little mystery. In the first place, it is easy to misperceive natural selection as a progressive process, to think that the fact that a species becomes better adapted to a specific environment means that it gets better in an absolute sense. But natural selection," says Gould, "is . . . a principle of local adaptation, not of general advance or progress."[13] Simply put, if the environment changes, previously maladapted variations may inherit the earth—for a while, that is, until the environment changes yet again.

Another reason for the illusion of progressiveness is human self-absorption. "Our impression that life evolves toward greater complexity," explains Stephen J. Gould, "is probably only a bias inspired by parochial focus on ourselves, and consequent overattention to complexifying creatures, while we ignore just as many lineages adapting equally well by becoming simpler."[14] No surprise, then, that Aristotle and Ptolemy placed the earth at the center of the universe, and that it wasn't until the sixteenth century that Copernicus began to convince people that the earth circled the sun. Not only does the Bible portray mankind as God's greatest creation, but virtually all religions presume a physical resemblance between man and God, and PBS commentator Bill Moyers says he was "stopped in his tracks" by a sculpture on the north portal of the cathedral at Chartres, which depicts Adam emerging "as an idea" from the side of God's head.[15] If, as Genesis instructs, it's all about us then, it is to be expected that we would portray ourselves atop the pinnacle of life at the end of a long upward slope of increasing improvement and sophistication.

But this slope is an illusion, or more precisely, a coincidence. It is true that life on earth began at a simple level, most likely as prokaryotes, single-cell organisms such as bacteria that lack a cell nucleus. Given the conditions on earth at the time, there probably was no other way that life could have begun. But as a result, all subsequent evolutionary developments had to consist either of similarly simple life forms or movement, at least temporarily, away from this "left-hand wall" of simplicity in the direction of greater complexity.[16] Perhaps under a different set of starting conditions, life might have emerged in a more complex initial state and subsequently have evolved toward greater simplicity. Indeed, there are numerous examples of decreasing complexity in evolution, such as the loss of functional eyes in cave-dwelling animals, simpler neural structures in salamander's brains than in their evolutionary ancestors, and the loss of body parts in parasites.

In short, there are a number of important unresolved questions about evolution: What is the primary evolutionary process, the fundamental unit of evolution? How fast does evolution take place? Can the environment change genes directly? How should living things be classified? And what does evolutionary theory tell us about man's place in the universe— is evolution progressive and man the supreme accomplishment of life, or is Stephen J. Gould right when he bluntly declares that "humans are here by the luck of the draw, not the inevitability of life's direction or evolution's mechanism"?[17]

One explanation for why there is so much uncertainty about matters so central to our understanding of human origins is the key-under-the-lamplight phenomenon. In a field in which there are a limited number of flickering lamps of knowledge and we can only see the sidewalk beneath where they shine, we are likely to think that we can find all the right answers there. Consequently, whenever a new piece of evidence turns up, whether it is a fossil or a new way of mining the information held by DNA, even the humblest researchers shout "Ah hah!" and proceed to try to convince everyone that they have found the key to the great puzzle.

These scientists are often respected members of their fields, and their stature alone helps overcome doubts about their claims. Not surprisingly, moreover, they do not take kindly to being challenged, and this leads to a noteworthy amount of pettiness and acrimony. One of the classic feuds is between paleontologists Martin Pickford and Richard Leakey. Leakey is the son of Mary and Louis Leakey, the famed fossil hunters who launched excavations at Kenya's Olduvai Gorge in the 1930s. Richard Leakey and Pickford had been friends when they were teenagers in Kenya but fell out as paleontologists. From 1968 until 1989, Leakey was director of the National Museums of Kenya. Before the Kenyan government would issue a permit to dig for fossils, researchers had to get permission from the museum, and Leakey was therefore in a position to prevent Pickford from excavating in the country, particularly in the area known as the Tugen Hills, in which both of them were interested. In 1985, Pickford visited the National Museums to take some notebooks written by his mentor, William Bishop, to be copied. Leakey accused Pickford of stealing the notebooks and banned him from the museum. Since this meant that Pickford could not get permission from the museum, he could not dig. So he joined up with a disgruntled former employee of the museum, Eustace Gitonga, who had started his own museum in Kenya, called the Community Museums, and got a permit from him.

In 1995, Pickford and Gitonga coauthored a book entitled *Richard E. Leakey: Master of Deceit.* In her 2007 book entitled *The First Human: The Race to Discover Our Earliest Ancestors*, Ann Gibbons, a correspondent for *Science* magazine, describes Pickford's book as follows: "The cover showed a drawing of Richard Leakey holding a skull with green dollar bills falling out of it. The title and cover were tame compared with the contents. . . . It started with a dedication to 'the victims of the fantastic Richard Leakey manipulations.' It described Leakey as a 'parasite,' his science as 'folly,' his books as a 'swarm of errors,' and his friends as 'toadies.'"[18]

Relying on the permit from Gitonga, Pickford continued to dig in the Tugen Hills. In 2000, at Leakey's instigation, Pickford was arrested for digging without proper permission.[19] He stayed in jail for several days before the case was dropped. Pickford and a French colleague then returned to the Tugen Hills and found the remains of a 6-million-year-old prehuman whom they named *Orrorin Tugenensis.* Given the timing in the year 2000, the fossil find was soon being called "Millennium Man." Meanwhile, Leakey accused Pickford of sending the fossils illegally to France for study.

Although the feud between Leakey and Pickford may be an extreme case, this backbiting isn't just a product of bile and dyspepsia. It happens to be one of the starkest examples of the principal of survival of the fittest in operation: the often brutal competition for academic tenure and professional stature. In paleontology, for example, graduate students and young professors make their careers by going on successful digs. The best success is a groundbreaking fossil find. Hence, there is pressure to deem as many fossils as possible to be groundbreaking. That is problematic because one person's momentous discovery is often somebody else's downfall.

Occasionally the new breakthrough merely replaces the old one as the talk of the town, which is punishing enough for the has-been. Frequently, however, the new discovery shows older claims to have been false. In Ethiopia in 1974, a professor of anthropology named Donald Johanson at Case Western Reserve University, which happens to be where I teach, and a graduate student named Tom Gray unearthed bones that they decided were the first hominid, thus making it the supposed missing link between chimpanzees and humans. Fossil hunters often give their skeletons nicknames, and Johanson's team called theirs Lucy after *Lucy in the Sky with Diamonds,* the song that they had played over and over again the night they celebrated their find.[20] Alas, in 1994 a group ironically also including faculty from Case Western Reserve University discovered the remains of a much older hominid that they named *Ardipithecus ramidus,* or Ardi for short. These scientists then spent the next fifteen years studying it, and in 2009, they announced that, like Lucy, slight, 4-foot tall Ardi was also a hominid, and that since she was more than a million years older than Lucy, Lucy wasn't the first hominid and therefore couldn't be the missing link. In fact, Ardi's discoverers declared, there *was* no missing link; it was supposed to be the connection between chimpanzees and humans, and Ardi showed instead that both humans and chimpanzees had branched off from a more distant common ancestor.

If Ardi's discoverers' interpretations were correct, she also negated the "savannah hypothesis" about why humans began walking on two feet instead of on all fours,[21] upon which an entire generation of evolutionary scientists has made its reputation. Paleontologists offer the savannah hypothesis as the answer to one of the most intriguing questions about human evolution: What characteristic first separated humans from apes? The answer to this question is important for human evolutionary engineering since it would help identify the human-nonhuman borderline that many say should not be crossed. The paleontologists' classic answer is bipedalism, walking on two legs. But why did humans begin to walk on two legs? The standard answer for that is the savannah hypothesis. As Swedish archeologist Bo Gräslund describes it, "During the Miocene, the period in which it was once believed that humanity evolved, a great drought turned large parts of the East African forest into savannah. Suddenly our forest-dwelling ancestors found themselves out there on the grasslands without any trees to climb. In order to survive they began to hunt, and in order to follow their prey and escape predators they raised themselves up on two legs and began to run."[22] Being bipedal, the theory goes, these early humans also could look for game over the high grass. Moreover, standing on their hind legs enabled them to stay cooler in the African sun. In Dean Falk's words, by standing upright, "early humans created their own shade."[23]

Once life on the savannah forced humans to become bipedal, the theory continues, they were able to begin to develop their remarkable cultural achievements: "In this way," explains Falk, "their arms and hands were freed up so that they could begin to make and use tools. This in turn stimulated the brain, which made further cultural achievements possible, leading in their turn to an even greater rate of intellectual development."[24] In short, the move to the savannah kick-started humanity's ascent up the evolutionary ladder. And on a somewhat less exalted level, the need to remain cool while running in the savannah sunlight also was said to account for some other notable physical differences between men and apes, namely, the absence of a fur coat and sweat glands, the latter of which, according to Dean Falk, gives humans "the greatest sweating capacity for a given surface area of any known animal."[25]

It's always a good idea to strengthen one theory by interweaving it with others, and to that end, the proponents of the savannah hypothesis proceeded to argue that humans became bipedal not only because, once on the grasslands of Africa, they had to run to hunt for food, but because once they caught and killed it, they had to haul it back to their homes on the edge of the forest. Being bipedal was selected for because it freed their arms and hands to carry or drag game. The savannah hypothesis thus is said to furnish a basis for a compelling three-way evolutionary connection: being bipedal, having large brains, and the development of culture. Our brains became larger because, compared with the types of food available previously, the game on the savannah provided more protein, which a larger brain required for nourishment; being bipedal enabled us to catch the game; and having free hands meant that the hands were free not only to carry the food home but to make better tools. An addendum to the savannah hypothesis also posits that, since the men were out hunting, the women could stay behind and take care of the home and the children.[26] Baby minding was especially important from an evolutionary perspective: human infants were vulnerable and feeble, thanks to having to be born very prematurely because their mother's pelvises couldn't grow wide enough to enable the babies' large heads, which contained their big brains, to pass through the birth canal without sacrificing the strength of the bony scaffolding necessary to support the mother's upright posture.

Despite its elegance, many if not most experts now believe that the savannah theory is incorrect. There are a number of reasons. First, the purported connection between bipedalism and freeing the hands to carry things is belied by the fact that apes and many monkeys carry things quite well despite being quadrupedal. The connection between being bipedal and taking care of helpless infants also is suspicious. As Gräslund points out, chimpanzee babies are similarly vulnerable for a long period of time after birth, but they manage to cling quite successfully to the fur on their quadrupedal mothers' backs. As far as Gräslund is concerned, in fact, "bipedalism did not arise to make it easier to carry small infants; it instead arose despite the fact that it made it *harder* to do so."[27] Gräslund also scoffs at the idea that humans became bipedal so that they could peer over the savannah grass. Men weren't the only hunters on the plain, he reminds us: "A person walking around at full height in open terrain with the idea of spotting dangerous predators will soon discover that the predators can see them as well. If our four-footed ancestors had had a problem with visibility on the savannah, they would have been much better advised to do the same as apes, monkeys, bears, prairie dogs, and many other animals, namely to raise themselves up on their back legs for a moment and then drop down on all fours again."[28]

As for the relationship between bipedalism and staying cool, it unfortunately ignores the fact that humans developed another feature that the apes lacked, an insulating layer of subcutaneous fat, which would make cooling ourselves more difficult. The proponents of the savannah theory respond that we needed the fat to keep our smooth skins warm, since the savannah, although scorching during the day, got cool at night. But this all begins to sound a bit contrived. Is it possible that we developed smooth skin in order not to get overheated when running on the savannah, subcutaneous fat in order to stay warm at night without fur, and sweat glands to avoid being overheated by the fat?[29] Columbia anthropologist Ralph Holloway sarcastically argues that "at this level of speculation, I could easily assert that as the Pliocene progressed and aridity increased, the distances between shade trees increased and hominins developed bipedalism so they could stand in the shade during midday more easily, thus reducing the risk of hyperthermia, inadvertently leaving the hands free to make sombreros."[30]

The savannah hypothesis also manages to offend an entire cadre of mostly female anthropologists who are insulted by the notion that women evolved merely to be housewives tending gardens behind the house while, in Falk's words, "men made the tools and brought home the bacon."[31] The offended anthropologists point out that these stay-at-home moms were responsible for the cultural breakthrough of farming, which, like hunting, was soon able to provide more calories than could be found in the wild to feed the swelling brain, and which eventually became the main source of all the calories in the human diet. What's more, the anthropologists add, it was the women who, being at home, had the time to fiddle around with those chunks of stone lying around. In short, the anthropologists note archly, "not only did early hominin women provide most of the food, but their freed forelimbs may have fashioned the first tools."[32]

The sharpest blow to the paleontologists' savannah theory, however, has come from their own discoveries.[33] Starting about a dozen years ago, fossils of bipedal humans began to be found in what clearly had been forest rather than savannah environments. Some paleontologists ventured that bipedalism had emerged on the border between forest and savannah, but this ran into the objection that modern wild chimpanzees who live in that same border zone behave entirely like forest chimps, rather than evolving a different set of behaviors suited to the neighboring grasslands. Moreover, the earliest bipedal skeletons were discovered without any tools nearby; tools only appeared in proximity to skeletons of more recent hominins. This showed that, contrary to the savannah hypothesis, humans had walked on two legs for several million years before they had begun to make stone tools. And then there was the discovery of Ardi, described earlier, a hominid who apparently walked upright and lived in the forest. (This of course raises the question of why bipedalism would have developed in the woods; one theory, called the aquatic hypothesis, claims that it is an adaptation for wading through forest pools.) For the present, the jury is still out on the explanation for bipedalism. As Dean Falk says, "bipedalism, more than anything else, separates early hominins from apes. Yet the primary cause for bipedalism remains the biggest unsolved mystery in hominin paleontology."[34] But the last word may well belong to Ernst Mayr, who observed that "the human species is highly successful even though it has not yet completed the transition from quadrupedal to bipedal life in all of its structures."[35] In other words, the real flaw in the savannah hypothesis is that, as many Americans who are afflicted with lower back pain no doubt can attest, we aren't really bipedal.

The experience of having one's theory, or in some cases one's life's work, overthrown, as may have happened to the supporters of the savannah theory, often leads established scholars to simply ignore evidence that conflicts with their positions. Some go even further, attempting, like Martin Pickford and Richard Leakey, to prevent rivals from gaining physical access to the fossil record. "To study fossils," notes Falk, "you've got to be *privy* to them. But . . . just a few people in the world control access to the archaeological sites and fossils that must be studied if one is to do the research necessary to understand our origins."[36]

Other beleaguered researchers resort to frontal attack. In May 2009, a 47-million-year-old fossil of a 9-month-old infant called Ida was unveiled as the oldest human ancestor at a lavish affair at the American Museum of Natural History in New York. New York mayor Michael Bloomberg was in attendance. The Norwegian scientist who had purchased the fossil from "private collectors" two years previously proclaimed it to be "a holy grail for paleontology" and predicted that "this fossil will probably be pictured in all the text books for the next 100 years."[37] Mayor Bloomberg, notwithstanding his lack of expertise in the field of evolution, dubbed it an "astonishing breakthrough."[38] Skeptics wasted no time leaping upon these claims, however, and quickly published a paper in the prestigious journal *Nature* in which they declared them to be spurious; Ida was not a member of the hominid family, they contended, but of a different family called adapids, which includes lemurs.[39] Similarly, when the island "hobbits" of Indonesia were discovered in 2003, critics insisted that the bones were not those of a parallel human species that had lived only 540 generations ago, but of a modern human with a pituitary deficiency.[40]

If the conflicts between fossil hunters seem bitter, imagine the hornets' nest stirred up when a rival group of researchers challenged the reliability of fossil evidence itself. This is occurring in connection with the hotly contested dispute over the origins of modern humans. Since the 1950s, everyone has been in agreement that Africa was our ancestral home—well, almost everyone; Dean Falk was maintaining in 2004 that "pinpointing the country or even the general region where hominins originated . . . [was] problematical,"[41] and Canadian anthropologist Pamela Willoughby still says that "the evidence relating to the African origins of modern humans produces more questions than answers."[42]

In any event, it seems clear that 100,000 years ago or so, the world was populated by several different types of hominids—*Homo erectus* in Asia, Neanderthals in Europe, and *Homo sapiens* in Africa—and that within 70,000 years, with the possible exception of isolated populations such as the "hobbits" mentioned earlier, there was only one hominid left, *Homo sapiens sapiens*. When did *Homo erectus* and Neanderthals get from Africa to Asia and Europe? The estimated date has been moved back in time from 1 million years ago, which was the prevailing view in 1994, to 2 million years ago more recently.[43] Were *Homo erectus*, Neanderthals, and *Homo sapiens* already noticeably different species when they left Africa, or did some earlier common ancestor emigrate and then evolve, more or less simultaneously, continents apart, into modern man? And there is still the main mystery: what happened to these other branches of the human evolutionary tree? We've already heard various theories about the Neanderthals, but what was the fate of *Homo erectus*?

One theory holds that when *Homo erectus* left Africa about 2 million years ago, he migrated to Europe and Asia and simultaneously evolved in a number of places at the same time into modern *Homo sapiens*. This theory, called the multiregional hypothesis, acknowledges that there were subsequent waves of human migration from Africa and contends that these newcomers absorbed the evolving populations they found throughout Europe and Asia by breeding with them. The opposing theory is called the replacement hypothesis. It maintains that *Homo sapiens* developed only in Africa, spread abroad relatively recently, perhaps only 60,000 years ago, and, when they encountered other hominids like *Homo erectus*, supplanted them by being better suited to the environment, rather than absorbing them through interbreeding.

The multiregional theory is the darling of the fossil hunters. One reason may be that it allows more of them to claim that the bones they find outside of Africa are from evolutionary lines leading directly to modern humans, which, as described earlier, makes the finds more impressive. That is not to say that they agree on all aspects of the theory; there are intense debates among them about how many migrations came out of Africa and when, as well as about how to interpret the fossil evidence in order to characterize the hominid populations found outside of Africa. A few even reject the multiregional theory altogether in favor of the replacement theory.[44] But most fossil seekers are adamant that the multiregional theory is basically valid and, above all, that the fossil evidence, upon which it is based, is the supreme source of knowledge about evolution. As University of Michigan paleoanthropologist Milford Wolpoff, a leading champion of the multiregional theory, and Australian colleague Alan Thorne bluntly put it, "The fossil record is the real evidence for human evolution."[45]

In the meantime, however, a different group of scientists began to attack the puzzle of human origins using evidence from their growing understanding of human genetics. Led by molecular biologists such as Berkeley's Allan C. Wilson, they searched for clues to human evolution in human DNA, and whenever the DNA evidence seemed to be inconsistent with the conclusions that the paleontologists had drawn from their fossils, the biologists not only disputed the paleontologists' claims but questioned the reliability of the fossil evidence upon which they were based. As Wilson and his Berkeley colleague Rebecca Cann put it, "fossil evidence is not objective, but rather, necessarily reflects the models the paleontologists wish to test,"[46] which is a nice way of saying that the paleontologists shape their interpretation of fossils to suit their pet theories, in particular, the multiregional theory of modern human origins. The molecular biologists already had one victory under their belts: "Arguing from their fossils," Cann and Wilson observe, "most paleontologists had claimed the evolutionary split between humans and the great apes occurred as long as 25 million years ago. We maintained human and ape genes were too similar for the schism to be more than a few million years old. After 15 years of disagreement, we won that argument when the paleontologists admitted we had been right and they had been wrong."[47] Now the biologists were convinced they were about to win the battle over the multiregional theory.

In 1987, Wilson, Cann, and a third researcher, Mark Stoneking, published an article in *Nature* entitled "Mitochondrial DNA and Human Evolution," in which they reported their analysis of human DNA found in the mitochondria, structures outside the cell nucleus that are inherited entirely from one's mother.[48] Their analysis convinced them that "all humans today can be traced along maternal lines of descent to a woman who lived about 200,000 years ago, probably in Africa."[49] The woman quickly became known as "Mitochondrial Eve." According to Wilson and his colleagues, she proved that "modern humans arose in one place and spread elsewhere."[50] In short, the multiregional theory was wrong.

The champions of the multiregional theory did not take this lying down. In a duel fought across the pages of *Scientific American* in 2003, Wolpoff and Thorne challenged every argument Wilson, Cann, and Stoneking were making in another article in the same issue of the magazine. In the first place, noted Wolpoff and Thorne, replacement theory has the newcomers from Africa being better adapted to the conditions they found in Europe and Asia than the native hominids who had been adapting there—according to most estimates, for more than 2 million years. This is inconsistent with the fundamental theory of evolution, which holds that populations become better adapted to stable environments over time. Nor is there any fossil evidence showing that the Africans brought with them any powerful, new technology that would give them a significant adaptive advantage. In his accompanying article, Wilson replies that the African humans did indeed bring a game changer with them: a mitochondrial gene for language lacking in the existing population.[51] According to Wilson, this gene enabled them to survive while the others died out not only because their language ability was so much better, but because they did not deign to breed with the others, thus both isolating them and outcompeting them. Wolpoff and Thorne retort that mitochondrial DNA cannot carry a gene that improves language ability because that would violate a condition that molecular biologists themselves impose on mitochondrial DNA in order for it to yield evolutionary information, namely, that mutations in mitochondrial DNA, as opposed to in the DNA found in cell nuclei, can have no positive or negative effect on fitness. And wouldn't you know it, they add, analysis of *nuclear* DNA supports the multiregional hypothesis.

But that's not all, add Wolpoff and Thorne: As you have disparaged our reliance on the fossil record, Professor Wilson, so we will discredit your reliance on mitochondrial DNA. For one thing, mitochondrial DNA does not take account of daughterless descendants; therefore, there could be plenty of sources of modern mitochondrial DNA other than Eve.[52] Furthermore, two of your own students (Stoneking, who coauthored with Wilson the 1987 paper in *Nature* that announced the discovery of Mitochondrial Eve, and Cann, the coauthor of the current piece in *Scientific American*) admit that their mitochondrial analysis can only place Eve somewhere between 50,000 and 500,000 years ago, rather than at 200,000 years ago, as you claim. In short, the mitochondrial DNA technique is not precise enough to support your date for Mitochondrial Eve.

In their 2003 article, Cann and Wilson also insisted that their DNA analysis showed that modern humans had replaced existing human species "*without any* genetic mixing" (their emphasis).[53] How this could happen, they admitted, "is still a compelling mystery," although they offered a few possibilities. But in 2010, as noted in the previous chapter, DNA researchers announced that Neanderthals and modern humans had in fact interbred, another blow to the replacement theory.

The disagreements among fossil hunters on the one hand and between fossil hunters and molecular biologists on the other involve only two of the many disciplines in which academics compete to solve the enigmas of evolution. In addition to paleontology and molecular biology, there's palaeoanthropology, biological anthropology, social anthropology, primatology, archaeology, zoology, linguistics, genetics, geography, taxonomy, phylogenetics, physical anthropology, neurology, developmental psychology, linguistics, archaeology, primatology, sociocultural anthropology, paleoclimatology, anatomy, physiology, and the list goes on. Within each of these disciplines, ambitious up-and-comers have an incentive to challenge established figures, and both rookies and veterans compete with their counterparts in the other subjects.

On top of that, new fields are constantly springing up or splitting off and proclaiming themselves to be the new truth. A recent example is evolutionary developmental biology, or "evo-devo" for short. One of its founders, Sean Carroll at the University of Wisconsin, describes it as "the study of the pivotal role played in evolution by genes and processes associated with the development of anatomy"[54] and claims that "much of what has been learned [from evo-devo], about animal forms in particular, has been so stunning and unexpected that it has profoundly expanded and reshaped the picture of how evolution works."[55] Shots like this, of course, are acts of war, and the rivals whose "picture of evolution" Carroll claims to have reshaped promptly mobilize their counterarguments.

With this much turmoil and animosity among those who seek to understand it, is it any wonder so much remains uncertain about human evolution? And we haven't even mentioned the creationists, happily throwing sand on the whole enterprise.

So where does this leave us? Scientifically, there is no question that natural evolution is real. Furthermore, we probably understand enough about it to be able to identify a number of potential threats to the continuation of our evolutionary lineage, such as a catastrophic loss of genetic diversity. But the lack of clarity about so much of the natural process of evolution complicates efforts to prevent evolutionary engineering from making harmful mistakes. For example, recall the concern about crossing the line between humans and other animals, as well as Leon Kass's insistence on preserving our "given nature."[56]

Has the science of human evolution brought us any closer to knowing what that gift is, or even whether it truly exists? Is it bipedalism? Would we no longer be human if we gave our descendants the ability to walk up walls using all fours? Do our brains make us different from other animals? If so, says Dean Falk, it is not the basic anatomy of the brain: "Although the cerebral cortex of humans is larger and more convoluted than that of chimpanzees, human and chimp brains appear superficially to have the same parts." Nor does it appear to be the size of our brains. Earlier hominins had larger brains but less technology, and Ernst Mayr is astonished by the fact that, despite its enormous cultural achievements, "the human brain seems not to have changed one single bit since the first appearance of *Homo sapiens*, some 150,000 years ago. It seems that in an enlarged, more complex society, a bigger brain is no longer rewarded by a reproductive advantage. It certainly shows that there is no teleological trend toward a steady brain increase in the hominid lineage."[57] If there are essential differences between our brains and those of other animals, says Falk, "it appears that subtle differences in the nervous systems of humans and chimpanzees, not gross anatomy, are what count. These involve, in particular, the wiring and the distribution of neurochemicals in the frontal lobes and the association cortices, as well as the right/left contrasts in brain organization called lateralization."[58] Does this mean that these subtle differences should be off-limits to evolutionary engineering? No one knows. Disciples of evo-devo argue that the only thing that really separates us from our closest relatives, the chimpanzees, is differences in genetic switches.[59] Can evolutionary engineering take place without altering those switches? If not, which ones are more critical than others in defining us as human beings?

In addition, our uncertainty about how natural evolution works and what future it would hold for us makes it difficult to predict whether future humans who developed along largely "natural" lines would be better off than those who were the products of evolutionary engineering. This cuts in two directions. On the one hand, it undermines opponents of evolutionary engineering who argue that we must allow natural evolution to take its course. For all we know, directed evolution would produce fitter future humans. On the other hand, our limited knowledge counsels us to be cautious so that we do not inadvertently blunder into an evolutionary minefield.

What should we do then? Should we err on the side of caution and try to stop genetic engineering from taking place? Or would that be falling into the trap that Sunstein contends is the flaw in the precautionary principle, namely, sacrificing untold benefit by being too cautious? On the other hand, if we permit evolutionary engineering to proceed, how do we minimize the risks? What actions would we need to take to prevent harm to children? Would those measures be sufficient to protect future generations and preserve the lineage? If not, what additional steps must be taken? How soon would those measures have to be in place? The subsequent chapters will explore the answers to these questions.

PART THREE

Managing Risk in Evolutionary Engineering

CHAPTER EIGHT

Protecting the Children

There are a number of ways in which evolutionary engineering could seriously harm those most immediately at risk, the children who were genetically modified. Children might be physically damaged by manipulations that went awry, a significant possibility given the complexities of human biology. Parents might make poor choices about how to modify their children that ended up compromising the children's welfare and, from an evolutionary standpoint, reducing the children's adaptive fitness. Child development might be stultified if parents made narrow or restricting modifications. The parent-child relationship could be disrupted if parents felt guilty about the negative consequences of genetic engineering, were disappointed by children who failed to take full advantage of their enhancements, tended to treat modified children chiefly as objects to be molded in the parents' images, or had impoverished the family in order to pay for the engineering. Sibling relations could suffer because parents modified children in different ways, or only modified some of them, perhaps because the parents couldn't afford to do more. Finally, the children could face crippling social discrimination.

Given these hazards, it is worth considering whether to ban evolutionary engineering altogether. There is some precedent for this. Australia, Canada, France, Germany, the Netherlands, South Africa, and Switzerland have enacted laws making it illegal to engage in human germ line genetic manipulation.[1] In 1997, the Council of Europe adopted a Convention on Human Rights and Biomedicine that only permits modification of human DNA for "preventive, diagnostic or therapeutic purposes and only if its aim is not to introduce any modification in the genome of any descendants."[2] In regard to the especially hot-button issue of chimeras, Australia, Canada, Denmark, Germany, and Spain have made it against the law to create certain types of animal-human hybrids.[3] Declaring that "human life is a gift from our creator, and that gift should never be discarded, devalued or put up for sale," George W. Bush urged Congress in his 2006 State of the Union address "to pass legislation to prohibit the most egregious abuses of medical research" and listed "creating human-animal hybrids" along with "human cloning in all its forms," as well as experiments on and commerce in human embryos.[4] (Bills to this effect have been introduced but not enacted.)

In 2008, the British Parliament passed the Human Fertilization and Embryology Act Amendments, which prohibit implanting into a woman a nonhuman embryo or reproductive cell or an "admixed" embryo.[5] However, the law allows the British Human Fertilization and Embryology Authority (HFEA) to grant licenses for researchers to create animal-human embryos so long as the embryos are destroyed before they are 14 days old.[6] Several months before the law was passed, the HFEA for the first time approved licenses for two experiments to insert a human nucleus into a cow egg, and while the bill was being debated, researchers at Newcastle University, which had received one of the licenses, succeeded in creating one of these "cytoplasmic hybrids," which survived for three days.[7] After the HFEA had issued licenses for three hybrid research projects in 2009, however, the British government refused to fund two of them, including the one ongoing at Newcastle.[8] By that time, the Newcastle researchers had produced 287 human-cow embryos.[9]

Despite its superficial attraction, a complete prohibition on evolutionary engineering would encounter a host of problems discussed in previous chapters. These include practical difficulties, such as how to stop black markets from developing; how to prevent parents from modifying their offspring in other countries where it might not be illegal; how to forbid objectionable uses of genetic engineering while permitting other uses that critics of evolutionary engineering might find acceptable, such as preventing the transmission of genetic diseases to one's children; and how to prevent the nation from losing its competitive edge against other countries that allowed or even encouraged citizens to engineer their descendants to be more inventive, more productive, more obedient, or better soldiers. Aside from these practical obstacles, a ban on all forms of evolutionary engineering would sweep up many practices that have become widespread and generally accepted, such as IVF and other assisted reproductive technologies for infertile couples, as well as pre-implantation testing of in-vitro-fertilized embryos to weed out those with serious diseases, not to mention medical care that treats or prevents illnesses that otherwise would have prevented their sufferers from having children. Even computerized dating services might be considered "assisted mating technologies" and therefore be fair game under a totally prohibitive legal regime.

But even if some forms of evolutionary engineering are permitted, what about germ line genetic engineering, in which active modifications are made to a person's DNA in such a way that they are transmitted via eggs and sperm to future generations? Isn't germ line engineering too dangerous? Shouldn't it be outlawed categorically, since it may be hard to determine if germ line changes would be safe for children, given that it may be necessary to wait for the child to reach adulthood and reproduce in order to detect latent adverse effects, and it may be even harder to feel confident that later generations will not suffer any serious problems? No biomedical intervention is completely risk-free, however; the standard that the FDA requires new drugs and medical devices to meet in order to be approved for marketing is not that they present no risk of harm, but that the risk is deemed acceptable in view of the potential benefits. In 2009, for example, the FDA Adverse Event Reporting System received almost 64,000 reports of patient deaths that were attributed at least in part to the use of prescription drugs under doctors' orders.[10] While some of these deaths may have been caused by medical malpractice, most of them undoubtedly were unavoidable consequences of proper physician behavior.

The question, then, is not whether germ line modifications are risky, but whether the potential benefits outweigh the risks, including the uncertain risks to modified children and their descendants. The response you get depends on whom you ask. In 2000, the AAAS issued a highly critical report on germ line gene therapy—the use of germ line engineering to cure, treat, or prevent disease. Many of the experts in their working group questioned whether there would ever be enough confidence in the safety of germ line therapy to justify its use.[11] This view was strongly influenced by the fact that the working group did not think that germ line therapy was particularly useful: "The working group could identify few instances where IGM [inheritable genetic modifications] would be needed. There are currently several alternative approaches available that will help parents avoid passing on defective genes to their offspring. These include genetic screening and counseling, prenatal diagnosis, and abortion, pre-implantation diagnosis and embryo selection, gamete donation, and adoption. In the future, *in utero* somatic gene therapy and gene therapy on patients after birth are likely to offer effective means for correcting defects."[12] The working group's argument rests on questionable assumptions, however. There are bound to be parents who prefer to have their own genetic offspring rather than to adopt or to employ surrogate eggs or sperm. Other parents may not want their children to have to undergo repeated somatic gene therapy treatments that could be prevented by germ line interventions performed when the children were early-stage embryos. It is difficult, therefore, to maintain that parents should be prevented from opting for germ line gene therapy categorically, under all circumstances.

But so far we have been discussing medical uses of germ line genetic modification, that is, to prevent deleterious genes from being passed on to one's children. The transhumanist vision described in chapter 1 goes much further, advocating the use of germ line engineering to improve nondisease characteristics such as intelligence, mood, and longevity. Even if we accepted medical uses of germ line engineering under some circumstances, shouldn't nonmedical enhancements be banned on the grounds that the risks to the children invariably would outweigh the benefits?

The argument that enhancement uses of genetic engineering should be prohibited while medical uses are permitted, however, rests on one and possibly two false assumptions. The first incorrect assumption is that medical benefits inherently are worth more than enhancement benefits, and therefore a degree of risk that would be acceptable in order to obtain a medical benefit would not be justified in order to obtain an enhancement benefit. This assumption has some superficial appeal; health is often described as a "special," "basic," or "primary" good,[13] without which it is impossible to attain "normal species functioning"[14] or to enjoy a reasonable range of opportunities.[15] However, the value of a medical versus an enhancement benefit clearly depends on the nature of the benefit. It is reasonable to expect that some medical benefits will be regarded as more valuable than some enhancement benefits. Most people probably would agree that a potentially life-saving drug or medical device provided greater benefit than a new nonprescription contact lens such as Bausch and Lomb's MAXSIGHT, which reduces glare for athletes. On the other hand, some enhancement benefits arguably are more valuable than medical benefits. An enhancement that increased cognitive function substantially is likely to be deemed more valuable than, say, a drug to treat nail fungus.

The second false assumption behind the argument that enhancement engineering ought to be prohibited even if medical engineering is not is that enhancement technologies are inherently riskier or that their risks are less well understood. The risks of gene insertion for therapeutic purposes, however, may be just as uncertain as the risks of gene insertion for enhancement purposes. Moreover, as noted earlier, essentially the same technical methods that are used to modify genes for health purposes will be used to modify genes for enhancement purposes; there is no obvious reason why they would work better for the former than for the latter. Finally, both medical interventions and enhancements may have both positive and negative social impacts. Therefore, it cannot be said that evolutionary engineering for enhancement purposes should be forbidden categorically.

A case can even be made for allowing the combining of animal and human DNA in certain cases. The research in Britain on inserting human cell nuclei into cow eggs that was cancelled for lack of funding was an effort to create stem cells from a patient's skin to treat illnesses such as heart disease, Parkinson's, and diabetes, thereby avoiding the controversial use of stem cells taken from human embryos.[16] This doesn't sound particularly horrifying, which is no doubt why the British parliament authorized such experiments and why the UK licensing authorities initially granted permission for them to take place.

What many people don't realize, in fact, is that humans already contain lots of nonhuman DNA. As one commentator explains, "at birth, your body was 100-percent human in terms of cells. At death, about 10-percent of the cells in your body will be human and the remaining 90-percent will be microorganisms."[17] The adult human body actually contains about 100 times more animal DNA than human DNA,[18] and as NIH director Francis Collins points out, if the nonhuman genes are added to our own, the number of genes at work in the human body increases from our native 23,000 or so to over 3 million.[19] The nonhuman DNA to which Collins is referring is found for the most part in bacteria that take up residence in the intestinal tract, where they perform critical functions in enabling us to digest our food, such as certain carbohydrates. One such microorganism is the E. coli that Paul Berg wanted to combine with a cancer-causing monkey virus when he was developing recombinant DNA. A recent study of 124 Europeans concluded that collectively their intestinal systems were inhabited by around a thousand species of bacteria, and that each individual harbored about 160 species.[20] Another report stated that there were as many as 10,000 different species of microorganisms and viruses in a single individual.[21] Humans could not survive without their colonies of these intestinal flora, which was recently affirmed when the NIH in 2007 established the Human Microbiome Project, which is aimed at studying the genetic makeup of these organisms and their role in human health and disease. The year before, in fact, a group of scientists had written in the journal *Science* that "humans are superorganisms whose metabolism represents an amalgamation of microbial and human attributes."[22] Instead of referring to the "human genome," they recommended that we use the term "metagenome."

Since we already host a swarming hive of symbiotic bacteria, why would it be abhorrent to supplement our resident bug population with some of their not-so-distant cousins? How about *Fibrobacter succinogenes* and *Ruminococcus albus*, the organisms that enable other animals to digest cellulose, the fibrous material that gives plants structural rigidity so that they can stand up straight? Recall that in Margaret Atwood's novel *Oryx and Crake*, Crake installs this capability in the Children of Crake, the new type of humans that he engineers to repopulate the Earth after he releases a catastrophic contagion. Cellulose salad isn't just the stuff of science fiction, moreover. According to Noah Schachtman, a reporter for the online technology journal *Wired*, the Department of Defense's Defense Advanced Research Projects Agency (DARPA) has funded a research project by U.S. Department of Agriculture microbiologists in Ames, Iowa, with the goal of identifying and studying the bacteria that secrete the enzymes that enable pigs to break down cellulose. Dubbed "Operation Intestinal Fortitude," the ultimate objective, according to Schachtman, is to develop "fibr-biotics" that will enable deployed soldiers to get more energy "from either food rations or non-traditional foodstuffs,"[23] that is, cellulose. Some major hurdles may stand in the way of success, though. For one thing, animals that digest cellulose such as cows and pigs have very different digestive systems than humans, including multiple and much larger stomachs. In addition, there is that rather noisome by-product of ruminant digestion, methane. It wouldn't do for soldiers creeping up on an enemy after a nutritious meal of leaves and twigs to give warning of the impending attack by their flatulence.

It might be objected, however, that when transhumanists propose giving humans advantageous animal capabilities, such as an eagle's eyesight, they have in mind adding animal genes to human DNA, while the nonhuman DNA in the human microbiome exists in nonhuman cells, such as those of bacteria, and has not been combined with the DNA in human cells. Therefore, the fact that the human microbiome contains massive amounts of nonhuman DNA does not necessarily justify incorporating animal and human DNA in the same cells. This is true, but the human genome actually does contain DNA from nonhuman origins. Approximately 8 percent of the human genome is composed of human endogenous retroviruses (HERVs),[24] which are remnants of infections that were produced in our ancestors by retroviruses, a type of virus that includes HIV and the virus that causes the common cold (and that killed Jessie Gelsinger). Over time, DNA from these retroviruses infiltrated and became part of the human DNA molecule. While the retroviral DNA in the human genome admittedly is not the result of the deliberate mixing of human and nonhuman DNA, as more people become aware that they contain nonhuman DNA, germ line engineering that intentionally mingles animal and human genes could lose much of its distastefulness.

The argument that all forms of evolutionary engineering should be banned in order to prevent harm to children, in short, is unsustainable. Humans have been intervening in the natural evolutionary process since the invention of medicine. The same widely accepted methods of assisted reproduction that are used to combat infertility and prevent the inheritance of genetic disorders will soon be able to be employed to genetically enhance offspring. Germ line genetic modification may be justifiable as a means of eradicating genetic diseases and disabilities from family lines. Even mixing human and animal DNA cannot be ruled out entirely. Instead of trying to block evolutionary engineering altogether, the focus instead must be on preventing unreasonable harm to engineered children. The question then is how much harm is reasonable.

Before seeking to answer this question, however, it is necessary to consider a rather startling contention, namely, that it is logically impossible for evolutionary engineering to cause *any* kind of harm to children. This contention is based on an argument made by Oxford philosopher Derek Parfit called "the Non-Identity Problem."[25] As applied to evolutionary engineering, the Non-Identity Problem would contend that if parents conceived a child for the purpose of engineering it and proceeded to do so in a harmful manner, we cannot say that the child has been harmed because if it had not been engineered, it would not have existed at all. Put another way, while a child who had not been harmfully engineered might be said to have a better life than the harmfully engineered child, they are not the same child, and therefore we cannot say that the parents have deprived the engineered child of a better life. To illustrate this argument, Parfit gives the example of a 14-year-old girl who is proposing to have a baby. We can advise her to wait until she is older on the premise that she will be better able to cope with the demands of parenthood, but we can't logically tell her that this would be better for *the baby*, since the baby she would have at 14 and the one she would have when she was older would be different people. Therefore, we cannot say that having the child at 14 is harmful to the child.

The Non-Identity Problem is more complicated, however. Technically, if parents conceived a "normal" embryo or fetus and would have allowed it to be born in any case but then proceeded to engineer it in a harmful way, we *can* accuse them of having harmed the child, since if the parents hadn't engineered it harmfully, the child would still have been born and would not have been harmed. This shows that a lot turns on the parents' intentions and on the precise sequence of events: if the harmful engineering occurs early on or as an integral part of the process of conception, for example, if the parents engineer their eggs, sperm, or an early-stage embryo, the resulting child cannot be said to have been harmed, but if the engineering occurs *after* the child is conceived, say, at a later stage of embryonic or fetal development, and the parents would have had the child even if it had not been engineered, then the child *can* be said to have suffered harm.

John Robertson, an American legal scholar who is a staunch defender of the right of parents to make reproductive choices, including the choice to genetically modify their children, makes this point in saying that "if the embryos that are altered would have been transferred [to the uterus] anyway, the alteration risks harms to offspring who could have been born without the risk of that harm."[26] On the other hand, Robertson observes, if parents would be willing to have children only if they were genetically modified to their taste, then the alternative to not being genetically modified is to not exist at all, and since existing is presumed to be preferable to not existing, parents ought to be allowed to genetically engineer their children in any way that they see fit. Philosopher Ronald Dworkin applies the same sort of reasoning to support IVF, which troubles pro-life advocates because, among other things, it ends up discarding most of the human embryos that it produces. "We accept in vitro fertilization as a reproductive technique," says Dworkin, "because we do not believe that it shows disrespect for the human life embodied in one zygote to allow it to perish when the process that both created and doomed it also produces a flourishing human life that would not otherwise have existed."[27]

The notion embodied in the Non-Identity Problem that a child cannot be harmed if the alternative is not to have been born at all has played a major role in judicial decisions in what are known as "wrongful life" cases. These are lawsuits that allege first that, owing to the defendant's negligence, a child was born who would not have been born if the defendant had not acted negligently; and second that, as a result of being born, the child has been harmed in some way. An example would be a couple that goes to a doctor for genetic testing to see if they are carriers of a certain genetic disorder and, upon being negligently misinformed by the doctor that they are not, proceeds to conceive a child who turns out to have the disorder.[28] If the doctor had correctly informed the parents, they would not have conceived the child and it would not have existed. Another illustration would be a couple that has one child who has cystic fibrosis and that relies on a fertility clinic to employ IVF and pre-implantation genetic testing to avoid having another child with the disease; if the fertility clinic makes a mistake and implants the wrong embryo, one that has the cystic fibrosis mutation and that would not have been brought to term except for the clinic's mistake, the resulting child would not have been born in the absence of the mistake.[29]

The parents in these cases may sue for their own damages, such as the extra costs of raising a child with a genetic disability compared with a normal child, and these suits by parents are called "wrongful birth" actions and are occasionally successful. But sometimes a claim is brought not on behalf of the parents for their losses, but on behalf of the child itself, seeking compensation for the child's losses—the extra medical and other expenses that the child will shoulder upon reaching adulthood, and the pain and suffering that the child claims to have suffered as a result of the defendant's negligent behavior. These suits on behalf of the children are called "wrongful life" actions and, for the most part, have not fared well in the courts because of the notion stemming from the Non-Identity Problem that if the defendants hadn't been negligent, the result would not have been the birth of a normal, healthy child but no child at all. In other words, the plaintiff child is arguing that he or she would have been better off never having been born. As the California Supreme Court put it in one case, *Turpin v. Sortini*, "it is simply impossible to determine in any rational or reasoned fashion whether the plaintiff has in fact suffered an injury in being born impaired rather than not being born."[30]

But there is another complication to the Non-Identity Problem. To say that someone who would not have been born except for someone's wrongdoing has nothing to complain about assumes that the person's quality of life is at least minimally acceptable. Otherwise, it may well be argued that, for this person, nonexistence *would have been better* than existence. The *Turpin* case just mentioned was brought on behalf of a child whose only impairment was being born deaf, which didn't strike the court as enough of an affliction that it would have been better if the child had never existed at all. But in an earlier case, *Curlender v. Bioscience*, which involved a child who was born with Tay-Sachs disease after her parents had been told erroneously that they were not carriers of the gene for the disease, a court reached the opposite conclusion. In contrast to the judge's reaction in the *Turpin* case to being born deaf, the judges who decided *Curlender* were struck by the devastating nature of Tay-Sachs, which a judge in another case described as follows:

> Tay-Sachs disease is a fatal genetic disorder that occurs in some children and causes the gradual degeneration of the central nervous system. While the child appears normal at birth, symptoms inevitably appear before the child reaches 8 1/2 months. At first the child is noticeably lethargic and development of his motor skills begins to decline. In addition, the Tay-Sachs child now becomes hypersensitive to noise. Between 12 and 24 months the child becomes blind, experiences petit mal seizures lasting for several seconds, is unable to eat because of the deterioration of his respiratory and digestive systems, and loses muscle strength. By the beginning of the third year, the child is blind, retarded, deaf, and completely paralyzed. By 40 months, most Tay-Sachs children will die of infections or other complications caused by Tay-Sachs.[31]

The court in the *Curlender* case agreed with this judge that if the defendant had not been negligent, the child would not have had to endure "such a short life of excruciating pain, devoid of any redeeming benefits."[32] The court in *Curlender* therefore allowed the child to recover damages for pain and suffering, asserting that the child "both *exists* and *suffers*, due to the negligence of others."[33]

So the Non-Identity Problem does not make it alright to engineer children in harmful ways. If the children's resulting quality of life is so poor that it can be said that they would have been better off if they had not existed at all, or if the children would have been born in a healthier state had the parents not engineered them the way they did, then the parents cannot escape responsibility.

Previous chapters described how parents might harm their children by modifying them in ways that left them grotesquely deformed, seriously ill, severely limited in their options, or treated like pariahs by the rest of society. But it is hard to imagine that parents would *intentionally* harm their children in these ways. Why would any parent engineer a child so that he or she was ugly? Why would a parent design a child to not be smart—so it wouldn't mind doing menial labor? If parents fashioned a child to excel at something for the parents' benefit, who's to say the child wouldn't benefit as well? Why is being enhanced in such a narrow way that a child is left little choice about what talents to cultivate worse than not being enhanced at all? If this narrow enhancement is all the parents can afford, is it better instead to leave the child unenhanced?

So the problem is not likely to be that parents deliberately engineer their children in ways that the parents know to be harmful, but instead that there is disagreement about what types of engineering are in fact harmful or that the parents do not engineer children carefully enough, acting too hastily or heedlessly and imposing too great a risk of injury on them. This leads to the question of how much leeway parents should be given to decide what is or is not harmful, and how much risk of harm parents can legitimately impose.

As a legal matter, the answer is "plenty." In terms of making health care decisions, for example, judges have upheld parents who refused to consent to corrective surgery for a heart defect,[34] withheld consent to chemotherapy,[35] denied permission for psychotropic drugs to be administered to their mentally ill children even when the parents no longer had custody,[36] and blocked a spinal tap to detect meningitis in a 5-week-old infant.[37] Albany Law School professor Alicia Ouellette notes how extensive parental authority is over medical interventions: "Parents have used their power to westernize the eyes of their adoptive Asian children, to modify the facial features of children with Down Syndrome, to inject human growth hormone (HGH) into healthy children, to enlarge the breasts of or suck the fat from teenagers, to attenuate the growth and remove reproductive organs of a child with disabilities, and to remove bone marrow from a nine-year-old girl for use by a brother who sexually abused her."[38] In my book *The Price of Perfection*, I described the severe risks of injury that parents are allowed to expose their children to in sports. As Ouellette concludes, "U.S. law allows parents extraordinary power over their children's bodies."[39]

What explains this deference to parental decision making? This can be partly attributed to tradition, stemming from the time when children were viewed as their parents'—or more precisely, their fathers'—property. From precolonial times until well into the nineteenth century, states family law expert Barbara Bennett Woodhouse, "the father's power over his household, like that of a God or King, was absolute."[40] Historian Michael Grossberg writes that the law treated children "as assets of estates in which fathers had a vested right. . . . Their services, earnings, and the like became the property of their paternal masters in exchange for life and maintenance."[41] Woodhouse observes that "the father was entitled to use the child as a productive asset to herd, spin, farm, or care for younger siblings."[42] This extended to discipline as well; fathers could beat their children (and their wives) with a stick so long as it was no thicker than a man's thumb,[43] which some have suggested was the origin of the phrase "rule of thumb," and in several colonies, children could be put to death for hitting and even for cursing their fathers.[44] The idea that children are the property of their parents has since been abandoned, leading to a different justification for parental hegemony—that parents usually are in the best position to do what's right for their children. As the Supreme Court stated in a 1979 case, the law historically has recognized "that natural bonds of affection lead parents to act in the best interests of their children."[45] This leads Dartmouth bioethicist Ronald M. Green to argue that parents should be given extensive leeway in deciding how to design their children: "Parents are best suited to understand and shape the lives of their offspring. Their freedom of decision in this area should have presumptive priority in our moral and legal thinking. Only in extreme cases are we warranted as a society in denying them access to the professional services they need to realize their choices or in preventing them from exercising those choices."[46]

Another rationale for deferring to the wishes of parents in designing their children is the parents' fundamental right to reproductive freedom, which the Supreme Court has recognized in a long line of cases, including *Roe v. Wade*. Since under *Roe* parents have the right to abort a fetus for whatever reason during the first trimester, it would seem to follow, for example, that they have the right to abort a fetus based on its non-disease characteristics. Furthermore, although the Supreme Court has not considered whether the rights associated with procreation extend to assisted reproductive techniques such as egg and sperm donation, IVF, and pre-implantation genetic testing, the increasing prevalence of these practices suggests that they too may deserve protection under the umbrella of reproductive freedom. In that case, any attempt by Congress or a state legislature to enact a law that interfered with such a fundamental right would have to pass an extremely stringent test; the courts would have to be persuaded that the law was the "least intrusive means" of furthering "a compelling state interest."

On the other hand, parents' decision-making authority over children clearly is not absolute. The state has intervened, for example, when parents failed to obtain treatment for a 5-week-old infant with two broken arms;[47] when a mother exposed a child to secondhand cigarette smoke during visitations;[48] and when a Norwegian-born mother tried treating her seriously burned daughter at home with wheat germ oil, goldenseal, comfrey, myrrh, and cold water, rather than taking her to a hospital.[49] Judges also routinely order blood transfusions and other medical interventions over parents' religious objections. As the Supreme Court has stated, "a state is not without constitutional control over parental discretion in dealing with children when their physical or mental health is jeopardized."[50]

So when is the government entitled to intervene in parental decisions affecting a child's health, including decisions about how to engineer children genetically? In some situations, courts have ordered parents to act in what the court thinks is in the child's "best interests," which might be deemed to authorize state interference whenever government officials felt that they knew better than parents what was best for a child. But applying the best interests standard to evolutionary engineering would risk a repeat of the eugenics movement, where those who thought they knew what genes were best for society used involuntary sterilization laws and human breeding programs in an effort to improve human stock. Furthermore, giving this much power to the government would be inconsistent with modern principles of parental freedom. It is not surprising, therefore, that courts have only invoked the best interests standard in disputes between parents, such as custody proceedings where both parents in a divorce seek child custody, where it serves in effect as a "tiebreaker." As Barbara Bennett Woodhouse notes, it is not employed as a decision-making standard in disputes between parents and "strangers," which would include state officials.[51]

The prevailing legal rule instead is much more respectful of parental decision-making autonomy: legally, parents can do anything they want to their children so long as their actions do not amount to "abuse" or "neglect." A federal law originally enacted in 1974 called the Child Abuse Prevention and Treatment Act, designed to improve child protection by the states, defines abuse and neglect as "at a minimum, any recent act or failure to act on the part of a parent or caretaker, which results in death, serious physical or emotional harm, sexual abuse or exploitation, or an act or failure to act which presents an imminent risk of serious harm."[52] Despite this federal law, what counts as abuse and neglect for the most part is a matter of state law, and although the definitions vary somewhat from one state to another, they generally include actions that are not relevant to evolutionary engineering, such as being cruel, using excessive punishment, or involving children in illegal drug activity or pornography. However, the abuse and neglect laws also typically forbid parents from creating a "substantial risk of serious physical harm" or acting so as to "seriously impair or retard the child's mental health or development."[53] State law typically does not define the terms "substantial risk," "serious physical harm," and "seriously impair or retard," but it is not too much of a stretch to imagine that some efforts at evolutionary engineering would be seen as abusive or neglectful.

One option, then, would simply be to use the abuse and neglect laws to protect children from harmful genetic engineering. In their current form, however, it is doubtful that these laws would be able to do the job. For example, it is doubtful that these laws apply to a child who is injured before it is born, or indeed, before it is implanted as a fetus in the womb, which is when the genetic engineering that is likely to be available to parents for the foreseeable future would take place. Laurence McCullough and Frank Cherbenak, for example, ask whether "we consider parents' freedom to engineer their children an issue of the rights of an actual child, in which case the child has full moral status" and would, for example, be protected under the abuse and neglect laws, "or the rights of an eventual or future child, in which case they do not?"[54] During the administration of George W. Bush, Congress passed a federal law known as the Unborn Victims of Violence Act that makes it a crime to kill or injure an unborn child, but its protection only extends to children "in utero,"[55] and thus it would not apply to genetic manipulations that took place prior to implantation in the uterus. Only one state, Louisiana, designates embryos as legal "persons," and even there, it is not clear that someone could be punished under the law for injuring or even destroying an embryo before it was implanted.[56] Prosecutors in some states have sought to punish women for substance abuse during pregnancy under state child abuse and neglect laws, but only one conviction, in South Carolina, has been upheld by the courts.[57] Moreover, considering embryos to be protected under the abuse and neglect laws might lead to restrictions on a woman's right to an abortion that would be inconsistent with current constitutional doctrine.

Another problem with applying the approach used in state abuse and neglect laws is that these laws only speak in terms of risk, and not in terms of a balancing of risks and benefits. Parents are prohibited from creating a "substantial risk of serious physical harm" to their children, but can a substantial risk be offset by the prospect of an even greater benefit? Suppose, for example, that a genetic modification carried with it a 10 percent risk of causing serious physical harm to a child but also had a 90 percent chance of conferring substantial benefit, such as making the child much smarter and more creative. Since a 10 percent risk of serious physical harm is unquestionably a "significant risk," would parents be guilty of abuse and neglect regardless of the potential benefit to the child? On the face of these laws, the answer would appear to be "yes."

Interestingly, the AMA has taken a position on genetic engineering that is pretty close to considering risks and ignoring benefits. In a noteworthy exercise in long-range thinking, the AMA's Council for Ethical and Judicial Affairs in 1994 deliberated about the circumstances, if any, in which it would be ethical for physicians to genetically enhance children. In the council's view, "efforts to enhance 'desirable' characteristics through the insertion of a modified or additional gene, or efforts to 'improve' complex human traits—the eugenic development of offspring— are contrary not only to the ethical tradition of medicine, but also to the egalitarian values of our society. Because of the potential for abuse, genetic manipulation to affect nondisease traits may never be acceptable and perhaps should never be pursued." The council went on to state that "if it is ever allowed, at least three conditions would have to be met before it could be deemed ethically acceptable: (1) there would have to be a clear and meaningful benefit to the person, (2) there would have to be no trade-off with other characteristics or traits, and (3) all citizens would have to have equal access to the genetic technology, irrespective of income or other socioeconomic characteristics."[58] The second condition— no trade-off with other characteristics or traits—seems to embody the risk-only approach suggested by the literal language of the abuse and neglect laws; a genetic manipulation would have to be all benefit, since any risk would entail a trade-off of some sort. Another example of a risk-only approach is the proposition put forward by Susann Baruch of the Genetics and Public Policy Center that "new [genetic engineering] technology must be shown to be no more risky than the normal process of conception and birth,"[59] regardless, it would appear, of any potential benefits.

And yet, in the belief that the benefits are worth it, parents routinely subject children to what arguably are substantial risks of serious harm without being accused by the authorities of abusing or neglecting them. The prime example is the benefit from sports activities. A recent study found that between 2001 and 2005, half a million children in the United States between 8 and 19 went to the emergency room for concussions, and that half of these injuries were sports related. Although participation in team sports declined between 1997 and 2007, the number of emergency room visits for 8- to 13-year-old kids with concussions sustained in team sports doubled, while they increased by more than 200 percent for kids aged 14 to 19.[60]

At elite levels, young children often continue to compete despite being hurt. In *The Price of Perfection*, I described the ordeal that some young female gymnasts might endure if they want a shot at the Olympics, as depicted in a program on CNN in 2003. The program followed a group of 7- to 11-year-old aspiring gymnasts in Pennsylvania. These girls practiced six days a week, 50 weeks a year. At the competition to select those who would go on to the national level, one 7-year-old completed her routine despite having torn ligaments and an ankle broken in two places. Later in the year she broke her other ankle, which may have cost her a shot at the Olympics. Another competitor landed on her head during practice, resulting in what she referred to as "a little bit of a concussion," and she later competed in the selection contest despite tendinitis, fever, and inflammation of the growth plate in her heel. Two of the children selected for Olympic grooming after the tryout ended up missing the U.S. championship competition as a result of injuries, one because she broke her hand during practice, the other because she broke her leg.

The parents of these children may have been indifferent to the injuries, they may have considered the kids as instruments for satisfying the parents' own need for fame and fortune, or they may have felt that the potential rewards of competing in and perhaps winning the Olympics were worth the harm that their children suffered. Yet while these sorts of parents are often criticized by child advocates, so far as I know none of them have been found unfit or liable for abuse and neglect on account of the sports risks that they permitted their children to experience. This strongly suggests that, as far as the law is concerned, risks are not the only factor to be taken into account.

Another reason to interpret the abuse and neglect standard to permit consideration of benefits as well as risks is that this is the approach that the law takes when parents make medically related decisions for their children. The legal rules governing the participation of children in medical experiments, for example, recognize that the children themselves are unlikely to be able to protect themselves in these situations, and the rules therefore establish a fairly rigorous system of oversight. The parents must give their permission, and the child is also asked to assent if he or she is mature enough. In addition, however, the investigators, the government in the case of certain types of studies, and a committee at the study location called the Institutional Review Board all must agree that the risks of being a subject are acceptable. If the rules took an approach that was indifferent to the potential benefits from the experiment, they might be interpreted to prohibit children from participating in studies that created more than a trivial risk of harm to them. Instead, the rules allow children to serve as subjects in more dangerous studies so long as the children stand to gain a direct benefit and the benefit is deemed to outweigh the risk.[61]

The law also weighs risks and benefits when parents make treatment decisions for their children. As stated earlier, no medical intervention is completely safe, and therefore there is always a risk that patients will suffer adverse effects, including results so severe that they would have been better off had they not been treated. Competent adults are allowed to judge for themselves whether the risks are outweighed by the potential benefits, and parents are permitted to make the same calculation on behalf of their children. Once again, parents are not confined to giving their children medical care that presents no serious risks; they can resort to interventions that are quite dangerous if they have the potential to confer sufficient benefit.

Relying on the actual abuse and neglect laws to protect children from inappropriate evolutionary engineering therefore is problematic because the laws may not extend to modifications made to early-stage embryos and because on their face the laws only consider risk and not offsetting benefit. Nevertheless, the substantial risk standard that they incorporate can serve as a good starting point for creating a new law that specifically addresses the hazards to children posed by evolutionary engineering. To begin with, the "substantial risk" standard says that parents can modify their children any way they wish so long as they do not create a substantial risk of harm to the child; a risk that is highly remote, or a risk of trivial harm, should be of no concern to anybody but the parents and the doctors and other health care professionals who assist them. (The law would expect the health care professionals to reduce even minor or remote risks as much as possible, however.) If the modification did pose a substantial risk of harm to the child, the parents should still be allowed to opt for it so long as the risk was outweighed by a sufficient amount of potential benefit. But this assumes that the risks and benefits are clearly understood and that there is agreement on whether the benefits outweigh the risks. How do we protect children from modifications where the ratio of risks and benefits is unclear or controversial? Yale law professor Bruce Ackerman takes the position that any genetic modifications should be allowed except those that would be considered "disadvantageous by every member of the generation already in existence."[62] In other words, there would have to be unanimous agreement that an installed trait was harmful before society would have the right to interfere with parental decisions.

A better rule might be to allow parents to engineer their children in any way that they wished except in ways that no reasonable parent would think provided a net benefit to the child. This rule differs from Ackerman's in two important respects. First, it considers matters from the unique perspective of parents. People who have never been parents may not adequately consider the parents' legal and ethical responsibilities to their children, as well as the delicate balance that parents try to maintain between being overly controlling and giving children too much leeway to make mistakes. Furthermore, an evolutionary intervention that might be appropriate for competent adults to choose for themselves might not be appropriate for their children, either because it was too risky or because it represented too much of a limitation on a child's future independence. A second difference between Ackerman's approach and the rule I am suggesting is that Ackerman would seem to require us to conduct empirical surveys of people's attitudes toward specific genetic modifications, since in theory one positive response would make a modification acceptable. The "no reasonable parent" standard on the other hand does not refer to the beliefs of actual persons, but instead to a hypothetical "reasonable parent" whom we all wished we had had or could be. Moreover, the "reasonable person" standard is very familiar to the law, which has used it successfully to help ordinary folks, usually members of a jury, figure out how they think a person ought to behave. Therefore, we end up with the following rule: *Parents should be allowed to genetically engineer a future child except in ways that no reasonable parent would choose or that would expose the child to a substantial risk of serious bodily or mental harm or impairment that is not outweighed by the potential benefit to the child.*

This rule has a number of features to commend it. It gives deference to parents' decisions about what types of children they want, which is consistent with the well-established principle that parents are entitled to a great deal of freedom when it comes to controlling their children's upbringing. In addition, the rule relies on two concepts, "reasonable" and "substantial," that, while incapable of precise definition, are commonly used in the law, and that by their very indeterminacy provide flexibility so that the rule can adapt to new engineering technology. After all, a change that no reasonable parent would attempt at the present time may become reasonable in the future after further research. Furthermore, the rule does not single out specific types of genetic engineering for special approbation. Nothing is categorically forbidden, neither germ line modification nor combining human and animal DNA. So long as a reasonable parent would use the technology and it does not present a significant risk of substantial harm to the child that is not outweighed by the benefit, then in terms of protecting the child, there is no need to interfere with the parents' choice. Of course, this is not to say that germ line modification or the creation of an animal-human hybrid should now be permissible; a good argument can be made that neither is safe enough to try at this time. But we can certainly envision a future in which they became sufficiently safe to be allowable. Finally, although the rule is highly deferential to parents, it does not give them the final say. Society reserves the right to overrule parental decisions when necessary to protect children from unreasonable harm.

One remaining question is what should count as "genetic engineering" so that it is covered by this rule. As chapter 3 explained, many forms of reproductive behavior can affect the genes of children and their descendants, including a number of techniques that are widely available now, such as courtship and mating decisions, genetic testing of prospective parents, the use of donor eggs and sperm, and the testing and selection of embryos fertilized in vitro. Any of these has the potential to cause harm to an eventual child. Persons with genetic mutations that produce Down syndrome or other physical or mental disabilities who mated with others who had the same mutations would increase the risk that their children would also have the mutation, compared with conceiving a child with someone who did not have it. Egg and sperm donors can have genetic disorders. Parents have implanted embryos after testing them to make sure they are suitable bone marrow donors and then donated bone marrow to siblings after they are born, a procedure that carries with it both health risks and discomfort. Pre-implantation genetic testing increasingly is being used for nonmedical purposes, such as selecting the sex of the embryos to be implanted, although the government agency in Britain that regulates fertility clinics will not allow parents to select their child's sex unless in order to prevent the transmission of a genetic disorder that is carried on the X or Y sex chromosome.[63] While all of these interventions have an impact on the genetic makeup of future generations, their impact is limited, since the choices they represent all involve existing genetic combinations. This provides a measure of comfort that the parents' decisions will not cause substantial physical harm to their offspring or deliberately modify them in bizarre ways. This is not the case, however, with genetic manipulations that actively alter children's genetic material, namely, the addition, deletion, or manipulation of DNA. It therefore seems appropriate to limit our discussion, and the rule recommended previously, to the latter forms of evolutionary engineering.

Establishing the proper rule by which to judge parental decisions and determining which engineering decisions it should cover, however, are only the first steps in protecting children from harm. The rule also must be capable of being enforced. This will not be easy. Once engineering technologies are available on the open market, controlling their use becomes very difficult. Therefore, it might seem sensible to try to prevent objectionable types of evolutionary engineering from being developed in the first place. But the government has relatively little control over basic scientific research. At the federal level, the government's main leverage is money: it can refuse to fund objectionable research, such as the federal ban on funding stem cell research in which human embryos are destroyed (the Dickey-Wicker Amendment) or NIH's embargo on funding human germ line or genetic enhancement experiments, mentioned earlier. However, while funding restrictions like these can slow advances in evolutionary engineering, they are unlikely to halt them indefinitely. Private industry, which now spends more than twice as much as the federal government on research and development, will step in if it anticipates that there is money to be made, as is likely to be true in this case.[64] In addition, state governments may provide research funding, as they did in connection with stem cell research banned by the Bush administration. Moreover, it is worth repeating that the same basic science that would make it possible to engineer the human future would be instrumental in conquering disease and disability. Not even the most extreme opponents of evolutionary engineering are likely to possess enough political clout to thwart medical progress, even assuming that they wanted to. So it is safe to assume that, one way or another, the basic technologies for engineering evolution will emerge. Protecting children therefore will have to depend on regulating how these technologies are developed and how they are used, rather than on deciding whether or not they are allowed to come into existence in the first place.

One way that parents employing evolutionary engineering could impose unreasonable risks of harm on children would be by allowing them to participate as research subjects in overly dangerous genetic engineering experiments. A fairly elaborate set of legal protections exists to protect children serving as subjects in biomedical research, but it is not clear that they would cover experiments on germ line genetic engineering, since in these experiments the initial subjects would not yet be children, but instead embryos, eggs, or sperm. Currently there are no federal rules in the United States concerning research on human eggs or sperm. A strong argument can be made that so long as research is restricted to these reproductive cells and they are not fertilized and allowed to become embryos, there is no need to provide legal protections for them; even the staunchest pro-life advocates stop short of conferring a right to protection on mere cells.

There also are no research protections for the men and women who donate eggs and sperm for purposes of experimentation. Although the focus here is on the need to protect the experimental subjects rather than their progenitors, it is worth pointing out that there have been calls for greater protections for donors, especially for women who donate their eggs. For example, although women are allowed to sell their eggs to fertility clinics and infertile couples, some object to compensating them for providing eggs for research, especially if the women are poor,[65] since they might be induced to ignore or downplay the health risks inherent in the donation process, such as ovarian hyperstimulation syndrome, which can be life threatening.[66] Accordingly, the NAS has issued guidelines that prohibit making payments to women who provide eggs for use in research, beyond reimbursing them for their medical expenses, travel, and lost wages,[67] and California and Massachusetts have made this prohibition a matter of state law.[68]

Not everyone agrees that women should be deprived of the freedom to sell their eggs, however. After all, since everyone else involved in the research enterprise can make money—the experimenters, the fertility clinics that harvest the eggs, and the companies that market the outcomes of successful experiments—why shouldn't the women who donate the eggs? The Ethics Committee of the State of New York's Empire State Stem Cell Board has therefore adopted a resolution that would allow researchers to compensate women who provided their eggs for research at the same rate that they would be paid if their eggs were to be used for reproductive purposes,[69] about $10,000 per extraction cycle.[70]

Experiments designed to ascertain the safety of evolutionary engineering cannot stop with the genetic manipulation of eggs or sperm, however. To be able to produce children with desired traits, the altered reproductive cells must be capable of being fertilized and then of dividing and maturing. As Johns Hopkins bioethicist Ruth Faden explains, "The only way to confirm that laboratory-created sperm or eggs are functional is to use them in the creation of embryos." In order to be certain that an experimental embryo would be able to develop into a child, the general rule is that they must survive at least until they reach what is known as the "blastocyst" stage, about 100 cells. "There's no way to get around that step," says Faden.[71]

Yet, there are surprisingly few rules governing research on human embryos. Creating embryos solely for research purposes, after which they would be frozen or destroyed, is illegal in eight states (Massachusetts, Michigan, Missouri, Montana, North Dakota, Oklahoma, Pennsylvania, and South Dakota[72]), but aside from the Dickey-Wicker Amendment mentioned earlier that prohibits the use of federal funds for such research, there are no federal restrictions. This laissez-faire attitude at the federal level is perplexing in view of the pro-life lobby's insistence that life begins at conception, but it may be attributable to the difficulty of determining precisely what protections, if any, embryos should receive. Some legal authorities, for example, regard embryos merely as property.[73] Beyond the federal government, guidelines for embryo research are also lax. Aside from prohibiting payments to cell donors, guidelines for stem cell research adopted by both the NAS and the University of Michigan merely require that the donors give their consent for the resulting blastocysts (early-stage embryos) to be used in research.[74] Perhaps the idea is that since embryos themselves are not entitled to very much protection, the people who donate the cells used to create them can provide what little protection is needed by exercising the right to withhold their consent to the use of the embryos in ethically inappropriate experiments. As long as the embryos do not survive very long, that may be fine. Recall that the British HFEA can even license researchers to create animal-human embryos so long as the embryos are destroyed before they are more than 14 days old.[75] Other countries have enacted more restrictive laws, however. Austria, Denmark, Germany, Spain, Finland, France, Greece, Ireland, Netherlands, and Portugal prohibit creating embryos specifically for research purposes.[76] Brazil only permits research on embryos that have been frozen for at least three years,[77] presumably because the length of cryostasis signifies a lack of interest in bringing them to term.

Determining the effects of genetic modifications on future children, however, requires research that goes beyond experiments in which the development of embryos is intentionally arrested before they develop too far. To assess the effects on actual children, there must be experiments in which the engineered embryos are implanted in the womb and allowed to develop further. Once an embryo is implanted into a woman's womb, however, it becomes a fetus, which alters its legal status significantly and triggers a number of restrictions on experimental use. These restrictions are contained in a set of federal regulations that were adopted following revelations in the press in 1972 about the U.S. Public Health Service syphilis experiment at Tuskegee. In that experiment, which began in 1932, 410 African American men in rural Alabama who had contracted syphilis were left untreated and followed in order to chart the course of the disease. The men were not told that they had syphilis, and more important, they were not given penicillin in the 1950s after it had been shown to be an effective cure. By the time the study was terminated, all but 72 of the subjects had died. In response to the furor that ensued when the press accounts appeared describing the experiments, Congress established a federal commission to devise legal rules for protecting human experimental subjects. The commission presented its findings in a document called the Belmont Report, named after the inn and conference center where the report was written, and when regulations embodying its recommendations were jointly adopted by 18 federal agencies, they became known as the "Common Rule."

The Common Rule requires experiments using human subjects to be approved by institutional review boards (IRBs) at the sites that conduct the research, mandates that subjects give their informed consent to participate in experiments, and prescribes that the risks to the subjects must be outweighed by the potential benefits from the study. The Common Rule also provides special protections for certain so-called vulnerable populations, one of which is fetuses. As the regulations now stand, fetuses may serve as research subjects, but if they are exposed to more than a minimal risk of harm, the study must hold out the prospect of direct benefit to the fetus.[78] In other words, the experiment cannot be aimed at generating knowledge that will be used to help others but will not benefit the experimental subjects themselves. In addition, parents must give their consent for fetuses to participate in studies, and pregnant women whose fetuses serve as subjects may not be given any inducement to terminate their pregnancies.[79]

To ensure that genetic modifications were made successfully, however, experimentation ultimately must extend past the cellular, embryonic, and even fetal stages to include the birth of live children. As noted earlier, the Common Rule establishes special protections for children used as experimental subjects that are similar to the protections for fetuses. Again, the parents must give their permission, but so must the children if they are mature enough, and if the study presents more than a small risk of harm, children can serve as subjects only if they stand to benefit directly or if a special federal panel finds that "the research presents a reasonable opportunity to further the understanding, prevention, or alleviation of a serious problem affecting the health or welfare of children."[80]

While the protections afforded by the Common Rule may seem substantial, there are serious concerns about how well they protect subjects. For example, a group of experts led by Ezekiel Emanuel, the head of bioethics at the Clinical Center of the NIH, states that "almost no one feels satisfied with protections for human participants in clinical research." These experts identify a number of problems, including that the process of IRB review is overly bureaucratic and inefficient.[81] IRBs also have been criticized for being overworked and understaffed and making an insufficient effort to monitor studies once they've started, while their members have been reproached for having conflicts of interest, lacking adequate training, and being inconsistent and nitpicky in their reviews.[82]

But even if IRBs and the rest of the current components of the Common Rule system worked flawlessly, their ability to protect children in experiments on evolutionary engineering would be limited, since the Common Rule contains no protections for reproductive cells or early-stage embryos that were altered experimentally *before* they became fetuses or children. New rules therefore are needed to cover experiments on reproductive cells and embryos, and these rules must go beyond the protections called for in NAS guidelines described earlier, since those guidelines are only intended to apply to embryos that are discarded after being used in research, and not to embryos that are implanted in the womb and brought to term. A starting point for these new rules could be the standard formulated earlier to protect children; employing this standard, parents would be permitted to experiment with reproductive cells or embryos except in ways that no reasonable parent would choose or that would expose a resulting child to a substantial risk of serious bodily or mental harm or impairment that was not outweighed by the potential benefit that the experiment offered the child.

At present, the federal protections for human subjects embodied by the Common Rule apply only in two types of experiments: those that are funded by the U.S. government, and those that are submitted to the FDA in support of an application for permission to market a drug or other biomedical product that the agency regulates. The FDA has asserted jurisdiction over human genetic manipulation, regarding it as falling within the definitions of both "drugs" and "biologics" in the Federal Food, Drug, and Cosmetic Act,[83] but the agency's regulatory authority on this subject has not yet been tested in the courts. The problem is that the FDA does not have the statutory authority to regulate "the practice of medicine,"[84] and genetic engineering might be considered to be more like a practice and less like a product under the agency's purview. If the FDA is deemed not to have the authority to regulate genetic engineering under the current law, then it will have no ability to regulate either government- or privately funded genetic engineering research on human subjects, including embryos, unless the law is changed. It is possible, though, that in order to reduce the risk of being sued by subjects claiming that they had been harmed by participating in unacceptably dangerous studies, private companies might design their genetic engineering experiments to fulfill the requirements of the Common Rule even if they were not required to do so by FDA law; when the NIH required gene experiments that it funded to be approved by its Recombinant DNA Advisory Committee, for example, private sponsors also submitted their studies to the committee for review.

Another factor in determining if the FDA will be able to regulate evolutionary engineering experiments funded by the private sector is whether the private sector will consider the development of evolutionary engineering technology to be sufficiently lucrative to justify the estimated $1.2 billion that it costs on average to develop a new biotech product,[85] including the costs of getting it through the FDA's regulatory process. At first blush, it might seem obvious that the demand for ways to enhance children's genes would be so great that private entrepreneurs would flock to satisfy it and that the profits they would reap would more than compensate them for their expenditures on research and development. But a company's ability to profit from a technology depends on whether it can prevent other companies from marketing the same technology without paying for the costs of research and development, and this typically requires that the first company have exclusive marketing rights to the product, namely, a patent.

The idea of private ownership of biological building blocks such as genes, however, is highly controversial. Nicholas Thompson, a fellow at the New American Foundation, writes that "gene patents are different from other patents. Biotech companies are not just putting fish genes in tomatoes to make them grow in cold weather; they're manipulating and owning parts of humans that have existed since well before the first hominid speared his first antelope. Furthermore, all of the research, from the machines used to sequence genes to the actual structure of the genome, is based in large part on a $15-billion investment by the public in the National Institute of Health's Human Genome Project." Thompson adds that "right now, the patent office is our only instrument for policing the gene industry, but the agency is unarmed and patrolling on foot."[86] In 2010, the federal courts joined the fray when a district court judge in Manhattan invalidated seven patents on the breast cancer genes BRCA1 and BRCA2 owned by Myriad Genetics on the grounds that the subject matter of the patents was an unpatentable "law of nature."[87] If patents on genes are not upheld by the courts, private industry would be far less likely to seek FDA approval for new genetic engineering technologies, and this would limit the agency's opportunities to regulate human engineering experiments.

᠊᠊ᢒ᠊᠊

Apart from the need to regulate human experimentation, an approach is also required for determining if sufficient research has taken place to make it appropriate to use evolutionary engineering technology outside of the experimental context. Gregory Stock recommends letting parents make this determination for themselves. "The law is a blunt instrument when it comes to the nuances of individual situations," observes Stock, "and it will seem ever more out of touch as our choices become subtler. Imagine having to seek permission for the timing of your child's birth, laying out your reasoning and pleading your case. The most even-handed committee would seem oppressive, because the process would be so intrusive. I suspect we would do better to rely on parental decisions, unless consensus exists that there is a likelihood of serious harm."[88]

The problem, of course, is precisely the risk of serious harm to engineered children that evolutionary engineering creates. Parents are unlikely to have the expertise to enable them to understand the risks and to make an informed choice about whether or not to impose them on their offspring. Most laypeople lack even a rudimentary understanding of basic scientific and medical issues. A professor at Northwestern University who surveys U.S. adults to determine their knowledge of science, for example, reports that "American adults in general do not understand what molecules are (other than that they are really small). Fewer than a third can identify DNA as a key to heredity. Only about 10 percent know what radiation is. One adult American in five thinks the Sun revolves around the Earth, an idea science had abandoned by the 17th century."[89] On top of their ignorance about science, people are too vulnerable to marketing ploys by unscrupulous providers. Remember 23andMe, the company mentioned in the introduction that sells 116 genetic tests over the Internet? The kit includes tests for traits such as "avoidance of errors," height, longevity, intelligence, pain sensitivity, "odor detection," and "breast-feeding and IQ," which might be of interest to parents seeking to decide which of a number of IVF-produced embryos to implant. Yet the company admits that only 30 of the 116 tests it sells have been shown to be valid, including none of the ones just listed.

An alternative is to rely on the geneticists, fertility specialists, and other health care professionals who provide the technical engineering to steer parents away from inappropriate modifications. This may provide parents with a source of expertise, but there is a serious risk that these professionals would act in their own self-interest to maximize their income rather than discourage parents from pursuing overly dangerous engineering objectives. According to the Center for Science and Genetics, the fertility industry, which is likely to provide most of the engineering services, is a huge business, generating more than $3 billion of revenue annually. A thriving trade in genetic engineering would swell these coffers even more.

There is clearly a need, therefore, for government regulation. The government already regulates genetic engineering of animals and plants. Before a manufacturer can create a new type of genetically engineered plant, it must first obtain a permit from the Department of Agriculture's Animal and Plant Health Inspection Service. (Certain crops with a long history of safe genetic engineering, such as corn, cotton, potatoes, tomatoes, and tobacco, can be modified after the agency has merely been notified.) To obtain a permit, the manufacturer must demonstrate that it knows the function of the new genetic material that it proposes to add to the plant genes and that the added material will not produce disease in plants or poisons, will not be used to manufacture drugs or industrial raw materials, and does not derive from a known human or animal pathogen. The FDA regulates genetically modified foods and animals whose genes have been modified to make them living drug factories. For example, the FDA determines whether a genetically modified food is equivalent in terms of nutritional value and safety (especially by not being allergenic) to the traditional food it supposedly improves upon. If it isn't, then the manufacturer must obtain FDA approval to market the food, and the agency can take a product off the market if it later turns out to be unsafe. Under the Animal Protection Act, the Department of Agriculture has the authority to quarantine or restrict the movement of genetically modified animals that pose health or safety risk to humans or other animals.[90] The EPA regulates plants that are genetically engineered to be pest resistant, such as the StarLink corn mentioned in chapter 3. A manufacturer wishing to engineer a plant must get an Experimental Use Permit before conducting field testing, and the EPA must be satisfied with the field test results before granting marketing approval.

In terms of human evolutionary engineering, the FDA rather than the Department of Agriculture or the EPA would be the logical regulatory agency. As discussed earlier in connection with the regulation of human experimentation, there is some question whether it has legal authority over these technologies, but if the agency's position is disputed by organized medicine, Congress could always pass a new law expanding the agency's authority to cover genetic engineering, although opposition from physician groups, who fiercely protect what they view as their professional prerogative to practice medicine without FDA interference, could make this politically difficult.

Even if the FDA's jurisdiction to regulate the safety of evolutionary engineering were clear, however, there would be no guarantee that it would do an adequate job of protecting children. Early on, in part in order to encourage the burgeoning U.S. biotech industry, the agency took the view that recombinant DNA and other genetic manufacturing techniques raised no special safety concerns distinct from other methods of production. As a result, the FDA has received a lot of criticism over its handling of genetically modified foods, or as critics call them, Frankenfoods. This calls into question whether the agency would be committed to adequately policing human genetic engineering.

Most recently, the agency was embroiled in a controversy over its plan to permit a company to market the first genetically modified animal for human consumption, a salmon called AquAdvantage. This fish grows twice as fast as a normal Atlantic salmon, owing to the addition of two foreign genes: one from the Chinook salmon which produces growth hormone, and the other from an eel-like fish called the ocean pout which acts as an antifreeze. Normal Chinook salmon stop producing growth hormone when the weather gets cold, but the pout gene enables the modified salmon to continue to produce the hormone all year long. As a consequence, fish can reach market size in 18 months instead of the usual three years.[91] Anti-Frankenfish forces are particularly miffed because the agency claims that it cannot require the salmon to bear a label identifying it as having been genetically engineered, and because the agency will not allow unmodified farmed fish to be labeled as being nonengineered either, since that would imply that the natural fish was healthier, which the agency says has not been shown to be the case.[92]

One Las Vegas restaurant chef thinks consumers might accept the modified salmon if it was less expensive. "There are people who will purchase it because we are in an economic hardship," he explained. "But what concerns me more is that this will open up a floodgate of people genetically modifying like proteins and other meat we are eating."[93] Other chefs think customers will reject it. One of them says the fish "sounds kind of weird," adding, "We have people ask all the time where the food is from. I think customers want to know where the cheese or the pork or the mushrooms are coming from. . . . I think with the big green movement over the past couple of years, they'd be against [genetically modified salmon]."[94] Other chefs raise broader concerns. "There is no way I would be interested in serving [genetically modified] salmon," says one chef from Oregon. "The eventual damage to the environment would be catastrophic. Scientists say they have sterilized the GMO [genetically modified organism] fish, but eventually one will adapt and destroy the natural process." Another chef concedes that "we need a sustainable way to farm-raise fish because the oceans cannot keep up with human consumption." Genetically engineered salmon, he says, is not the answer, however. "As soon as the government allows a corporation to patent a method of raising fish, there is a certain stranglehold on the fish supply in the future."[95] The manager of one fish market in Chicago is suspicious of the underlying science. "How can they grow that much faster with less feed?" he asks. "I don't think I could trust something like that."[96]

In part because of the FDA's relaxed stance toward genetically modified food, Frances Fukuyama and Franco Furger propose that genetic engineering and new reproductive technologies should be regulated by an independent federal commission whose members would be appointed by the president and confirmed by the Senate.[97] Their proposed commission resembles the bioethics advisory committees appointed by Presidents Clinton, Bush, and Obama, but with a regulatory rather than merely an advisory function. Fukuyama and Furger take pains to insist that their commission would be immune to "the political and administrative pitfalls into which regulation on controversial matters easily falls," but they fail to explain how they would accomplish this. On the contrary, their call for presidential appointment and Senate confirmation would ensure that the members of the commission were political appointees whose beliefs conformed to the prevailing ideology within the Beltway, the way President Bush's Council on Bioethics, of which Fukuyama was a member and Leon Kass the chairman, reflected conservative religious beliefs. Indeed, Fukuyama and Furger seek to perpetuate their conservative bias by insisting that certain technologies should be banned from the outset, urging Congress to prohibit "reproductive cloning, germline genetic modifications, and certain forms of human-animal chimeras and hybrids."[98] A new bureaucratic entity to regulate evolutionary engineering may indeed do a better job than the FDA, but Fukuyama and Furger's is likely to be unduly politicized and inordinately restrictive.

Whether genetic engineering is regulated by a new body or the FDA, a key question is what kind of experimental data should be required to show that a new step in evolutionary engineering would not place children at an undue risk of harm. Some critics go so far as to claim that adequate information is impossible to obtain. "[B]iological characteristics or traits usually depend on the interactions among many genes," argues the Council for Responsible Genetics, "and these genes are themselves affected by processes that occur both inside the organism and in its surroundings. This means that scientists cannot predict the full effect that any gene modification will have on the traits of people or other organisms."[99] The AAAS is not quite that pessimistic but nevertheless insists on the need for data demonstrating that "an altered embryo [is] able to transit all human developments without a mishap due to the induced intervention. And for those techniques that add foreign material, there must be multigenerational data showing that the modification or improvement of a specific genetically determined trait is stable and effective and does not interfere with the functioning of other genes."[100] Lest this statement be interpreted to mean that the AAAS thinks this information will be readily obtainable, the AAAS report observes that "many members of the working group, including several of the scientists, question whether we will ever have enough confidence in the safety of IGM [inheritable genetic modifications] to proceed to clinical use. This assessment led some to conclude that it would never be scientifically and ethically appropriate to begin human applications until we can surmount this problem."[101] Scientists need to work out what testing would be able to predict the effects of genetic engineering on children. But there is no reason to believe that this technical problem is any more insurmountable than the other hurdles that evolutionary engineers face. The position espoused by the working group members whom the AAAS statement refers to is too categorical to be considered sound public policy.

But the nagging problem remains: how do we prevent parents from jumping the gun and exposing children to unreasonable harm before the risks of evolutionary engineering are sufficiently well understood and appropriately minimized? The problem is complicated by the realization that some parents, such as fervent transhumanists, may be far less reluctant than others to pursue dangerous engineering options. One of the direst threats to children may be from cults. Cults have a history of prolonged conflict with the authorities over how cult members treat their children. The Branch Davidians, the group whose compound in Waco, Texas, was raided by federal agents in a violent confrontation in 1993, first came to the attention of law enforcement when they were suspected of illegally selling automatic weapons, but what prompted the raid itself were allegations that the leader of the cult, David Koresh, was having sex with underage girls. The allegations turned out to be true, with the sex taking place with the consent of the children's parents. (Koresh is said to have justified his actions on the basis that his children would rule the world following the "end time," as foretold in Revelations.[102]) Fundamentalist Mormons, in addition to running afoul of laws against polygamy, have been accused of statutory rape for marrying and having children with young girls.[103]

Nor is sexual abuse the only type of harm that the children of cults suffer. In a bizarre incident in 2002, Jacques Robidoux was convicted of first-degree murder for allowing his 1-year-old child Samuel to starve to death. As related by Catherine Wessinger, a professor of the history of religion at Loyola University in New Orleans, "a sister of Jacques Robidoux who was stout and plain had received a revelation when Samuel was 10 months old that [his mother] Karen, who was slim and pretty, should only breastfeed Samuel as penance for her vanity. . . . Karen was told that she was ordered by God to consume only almond milk, nurse Samuel for 20 minutes every hour around the clock, and sing hymns praising God while she was nursing Samuel. Under these circumstances, Karen did not have sufficient breast milk to nourish Samuel, and he wasted away over 51 days."[104] And who can forget hearing the news in 1978 of the deaths of 276 children in the mass suicide of members of Jim Jones's Peoples Temple in Guyana?

Cults might very well pressure or persuade their members to engage in dangerous engineering experiments. Consider the Raelians, founded by French former race car driver Claude Vorilhon, who now goes by the name of Rael. The Raelians believe that life on earth started from DNA brought here by aliens called the Elohim, the Hebrew word for "Gods." Rael claims that his enlightenment began in 1973 when he was 27: "I was a journalist in France. I was on my way to my office at about 9 a.m. I felt pushed to go with my car to a volcano about 10 kilometers from the city. I left my car and went by the foot of a crater, then I saw in the sky a very shiny, flashing light. And a craft . . . appeared. Then one of the Elohim came out and that was the beginning."[105] An Australian newspaper reporter picks up the story:

A human being with a goatee beard, almond-shaped eyes, olive skin and a halo disembarked from the vessel, took Rael inside and explained, in perfect French, that the Elohim (a Hebrew term used in the Bible that means "people from the sky") [*sic*] had created Earth and its inhabitants from scratch, around 25,000 years ago. The Elohim then whisked him off to their planet, where he was reunited with his dead mother (cloned from a photograph) and met Buddha, Muhammad and Jesus. . . . Before leaving the Elohim's planet, Rael had sex with six biological robots. (Oh, and Christ, by the way, was revealed to be his half-brother, both having been fathered by the elder alien, Yahweh—hence what modern-day Catholics refer to as the Virgin Birth.)[106]

In 1978 and again in 2003, the Raelians claimed to have successfully cloned a human being, an assertion they refused to substantiate after being asked to produce mother and child for DNA testing. Rael also has expressed his approval of "designer babies."[107] Therefore, it isn't hard to imagine Raelians encouraging their members to genetically engineer their children, without necessarily waiting until the technology had become safe enough. Similar cults might spring up in the future, like the "Elohimites," the Raelian takeoff in Michel Houellebecq's novel *The Possibility of an Island* which eventually replaces humans with cloned "neohuman" ciphers.

If children in cults are endangered by evolutionary engineering, the law is relatively clear that the state has the authority to intervene to protect them, regardless of objections that the government is interfering with the members' free exercise of religion. "The Raelians, of course, can believe whatever they want to," observe George Annas, Lori Andrews, and Rosario M. Isasi. "But just as human sacrifice is illegal, experiments that pose a significant danger to women and children can also be outlawed, and the religious beliefs of this cult do not provide a sufficient justification to refrain from outlawing cloning."[108] The U.S. Supreme Court has affirmed, for example, that "the right to practice religion does not include liberty to expose the child . . . to ill health or death."[109] Protecting children in cults presents practical challenges, however. Cults are secretive, and law enforcement officials therefore may not learn what is happening within them until it is too late, as illustrated by the deaths of the Peoples Temple children. In addition, when the state moves to protect children in cults from physical harm, the usual approach is to take the children away and punish the parents. When they assaulted the Waco compound, for instance, the federal agents who suspected that the Branch Davidians were abusing their children intended to rescue the children and arrest the adults, although they unfortunately ended up killing over 80 people, many of whom were the very children they had intended to protect. In some states, parents convicted of child abuse and neglect face up to 30 years imprisonment. In 2008, almost 270,000 children were removed from their homes on the basis of child abuse and neglect, and 20 percent of them were placed in foster care.[110]

The rationale for separating the children from their parents is to prevent further harm to the children, but outside of situations such as cults where parents' freedom may be compromised, their judgment corrupted, and child abuse condoned or ritualized, it may not make sense to remove the children and fine or jail the parents. There might be little reason to expect that parents who had harmed their children inadvertently in a sincere effort to give them a better life through genetic engineering would continue to cause them injury after they were born. Taking children away in these circumstances or imprisoning these parents would cause further harm to the children, while fining the parents would deprive the children of parental resources.

A solution suggested by some legal scholars, such as Kirsten Rabe Smolensky, is that children should be allowed to sue their parents for the injuries suffered by the parents' excessive engineering enthusiasm. Smolensky argues that "a born-alive child harmed by direct genetic interventions should be able to sue his parents successfully for battery where the parents intentionally engage in a process that is substantially certain to make a harmful or offensive contact with the embryo, and to cause legal harm to the later-born child."[111] In response to the objection that these suits could end up harming the child financially, Smolensky replies that wealthy families would be able to afford the court costs and legal fees, some court awards would be covered by insurance, and in the remainder of cases, parents should only be required to pay modest or nominal damages, so that the child at least would have the satisfaction of having had its day in court. But could parents and children be legal adversaries without destroying the parent-child relationship? Like the option of removing engineered children from their family, Smolensky's proposal has a similar tendency to backfire.

Since penalizing parents, depriving them of their children, or allowing their children to sue them all seem ill-advised, we need to pick another target for our enforcement efforts in seeking to protect children from unconscionable evolutionary engineering. The prospect of do-it-yourself genetic engineering that would allow parents to bypass professional assistance seems highly remote, and therefore the logical regulatory targets are the physicians and other health care professionals who supply the parents with the engineering know-how that leads to the harm, as well as the institutions and industries that supply the associated products and services. The legal system already has all the machinery it needs to do the job. To begin with, since 1946, children have been able to sue physicians and others who negligently cause them to be born with injuries or disabilities.[112] (Prior to 1946, the courts had held that unborn children were legally indistinguishable from their mothers and therefore could not sue on their own behalf.) Initially, these prenatal harm actions were allowed to be brought only by children who had been viable fetuses at the time that the physician had caused their injury, on the theory that it was too difficult to prove what caused harm to fetuses before they reached that stage of development. Today, most courts allow suits by injured children regardless of whether or not they were viable fetuses at the time the negligent acts occurred. Of course, minors cannot bring lawsuits themselves, so their parents would have to sue on their children's behalf, but the parents' consent or active involvement in obtaining the engineering services should not insulate the professional from a suit brought for the injury to the child. Health care professionals even might be held liable for aiding and abetting parental abuse and neglect.[113] Civil suits against professionals also would encourage professional groups such as the AMA and the American College of Medical Genetics to adopt appropriate ethical standards for evolutionary engineering provided by their members.

There is one important difference, however, between run-of-the-mill medical malpractice and the liability of health care professionals for harming children by providing excessively dangerous engineering services. Ordinarily, physicians and other health care givers are judged by the standard of their profession, that is, they are expected to behave the way a reasonable professional would behave under the same circumstances. This standard of care typically is established through expert witnesses, who testify about how they believe a reasonable professional ought to behave. But earlier we said that, in terms of its effects on engineered children, the standard by which evolutionary engineering should be judged was from the vantage point of the parents, with engineering deemed acceptable unless no reasonable parent would choose it. At first it might seem unworkable to hold health care professionals to a parent-oriented standard of behavior; after all, how are geneticists supposed to know what choices reasonable parents would make? But there is precedent in the form of a similar approach that the law takes when it assesses whether or not a physician has obtained proper informed consent from a patient: by and large, physicians are not judged on the basis of how reasonable physicians would behave, but on the basis of how reasonable patients would want them to behave. Expert evidence is not required; instead, the jury (or judge, if the case is being heard without a jury) must determine what information a reasonable patient in the patient's circumstances would want to be given. Similarly, judges and juries ought to be able to ascertain what forms of evolutionary engineering reasonable parents would and would not regard as appropriate.

In addition to facing lawsuits by injured children, physicians and other licensed professionals who engaged in improper genetic engineering could face disciplinary action by state medical boards or other official bodies, including having their licenses revoked. Sanctions also can be levied by professional organizations such as the AMA for violating the organization's ethical codes, including revoking membership in the organization. Doctors who lose their membership in the AMA and similar professional organizations lose more than the monthly magazine or Hertz rental car discounts; it can cost them hospital privileges, increase their malpractice insurance premiums, and lead to disciplinary action by the state.

Children also can be protected by allowing them to sue not only the health care professionals who injure them but the fertility clinics and other institutions where these professionals work. This is similar to the manner in which injured patients can hold hospitals accountable for their own wrongdoing in failing to make sure that the doctors on their medical staffs are properly qualified and adequately supervised, an approach known as "hospital corporate liability." Ideally injured children also should be able to recover damages from the companies that manufacture the drugs and other products that health care professionals use in the course of the engineering process; however, industry pressure has succeeded in making it increasingly difficult to successfully bring product liability suits like these, and therefore professionals and institutional service providers are likely to remain the primary liability targets.

Finally, we also need to deal with the risk that children will be harmed because of state-sponsored, global genetic competition. There is no way to stop this competition, any more than we can stop the development of the technology itself, but we should push for an international treaty banning unethical, state-sponsored military or other genetic manipulation, the same way we have outlawed torture and poison gas. Note that this is not the same treaty proposed in 2001 by George Annas, mentioned earlier, which would have banned genetic engineering altogether.

Holding health care professionals and their institutions responsible for harming children by furnishing parents with unacceptably dangerous types of evolutionary engineering, plus adopting and enforcing an international agreement against harmful state-sponsored forms of engineering, would help protect children. But what about the risk that evolutionary engineering would lead to social strife and political upheaval? Would the measures that were taken to protect children suffice to protect society, or are additional steps necessary?

CHAPTER NINE

Preserving Societal Cohesion

It is possible that access to advantageous evolutionary engineering only by those who already enjoyed wealth and social privilege could fracture society into warring classes and provoke a rebellion by the un-engineered who felt that they no longer had an equal opportunity to obtain social rewards. As Sharon Beder writes in *Selling the Work Ethic: From Puritan Pulpit to Corporate PR*, "America's reputation as a land of opportunity rested on its claim that the destruction of hereditary obstacles to advancement had created conditions in which social mobility depended on individual initiative alone."[1] Parents whose children had not been genetically modified might insist on outing children who had and depriving them of any societal benefits that they might have been able to acquire through the use of their superior abilities.

Such a witch hunt would resemble the campaign against athletes who use performance-enhancing drugs, which is fueled in large part by the conviction that these athletes do not deserve to benefit from their accomplishments. For example, President George W. Bush's bioethics advisory council, headed by Leon Kass, accused athletes who used steroids of "getting their achievements 'on the cheap,' performing deeds that *appear* to be, but that are not *in truth*, wholly their own."[2] When Bush's attorney general John Ashcroft brought federal indictments in 2004 against a coach, a trainer, and two executives of a company named BALCO for distributing illegal steroids, he blamed steroids for "foster[ing] the lie that excellence can be bought rather than earned."[3] And in his 2004 State of the Union address, Bush branded steroids as "shortcuts to accomplishment."[4] (The irony, of course, is that Yale would never have admitted George Bush as an undergraduate on the basis of his own accomplishments. As Berkeley sociologist Jerome Karabel writes, while Bush had attended Phillips Academy, an exclusive prep school, he "had never made the honor roll and his verbal score on the SAT was a mediocre 566. Although popular among his classmates, he was neither an exceptional athlete nor did he possess any particularly outstanding extracurricular talents. . . . [But] as the son of a prominent Texas oilman then running for the United States Senate—and the grandson of a United States senator from Connecticut who had recently served as a member of the Yale Corporation, George W. Bush was no ordinary applicant."[5])

How might the genetic underclass wage a campaign to clip the wings of persons who had been genetically modified? They might try to ban genetic engineering, much as professional sports organizations have tried to block the use of performance-enhancing drugs. But as discussed earlier, it would not be easy to stop people from obtaining illegal technologies on the black market, or traveling abroad and securing them in countries that were more hospitable to the genetic engineering industry or simply more desperate for foreign exchange.

An alternative to prohibiting genetic engineering would be to try to level the playing field. Taking a lesson from athletics, for example, competitions might be forbidden between engineered and normal people. This already happens to some extent in sports: boxers and wrestlers typically compete only with others in the same weight range. Or, enhanced individuals could be allowed to compete against unenhanced persons, but only after being handicapped so that they lost their advantage. This also takes place today in sports: in "handicapped" horse races, jockeys who weigh too little must carry weights in their saddles, and strokes are subtracted from the scores of golfers who aren't as good as the other players in their group.

Any effort to create a level playing field, however, would necessitate being able to tell who had been genetically engineered. Since the engineered themselves would have no incentive to volunteer this information, there would have to be tests that enabled them to be discovered. One approach would be to require altered or added DNA to be tagged in some way that could be detected by a simple scan. But the benefits of being able to thwart the scanner would be so great that parents would seek out U.S. or foreign physicians or geneticists who, at the right price, might be willing to leave off the tag when they engineered embryonic DNA. Another option would be to obtain DNA samples from newborns and compare them with DNA taken from their parents to find differences that could only occur as a result of deliberate manipulation. Part of this scheme is already in place: by state law, blood samples containing DNA must be taken at birth in order to screen infants for disease. Yet again, determined parents might find ways around this, such as by hiring cooperative obstetricians or midwives who would be willing to substitute the parents' own blood or otherwise fake the samples, similar to how Vincent fools the scanners in the film *Gattaca*.

More important, even assuming that genetically engineered newborns could be identified at birth, how would they be identified later on in life? Could they be forced to carry some sort of ID? Couldn't the ID be counterfeited? Could they be branded in some indelible way? More likely, genetic tests that could reveal whether or not people were engineered would have to be employed before they engaged in competitions in which they would be deemed to have an unfair advantage. But the competitions would not just be athletic contests: a society intent on denying an advantage to engineered individuals would probably want to identify them prior to offering them employment or evaluating their job performance, before they sat for school tests and college entrance exams, in advance of delicate business negotiations, and so on. Again, there is an analogy to this in the testing conducted by anti-doping forces in sports, and while this might seem overly intrusive, genetic testing can be performed on DNA obtained by merely swabbing the inside of the cheek with a Q-Tip, making it much less intrusive than the blood or urine sampling to which athletes are subjected.

Assuming that engineered people could be detected, how would unfairness be prevented? Would they be barred from competing—getting jobs, for instance? If so, then society would lose all the benefit of their special capacities in the workplace. An alternative would be to handicap them in some way: deduct points from their test scores, make them work under less favorable conditions, or give them sedatives to blunt their negotiating skills. Kurt Vonnegut satirized efforts like these to level the playing field in his classic short story *Harrison Bergeron*: "The year was 2081, and everybody was finally equal. They weren't only equal before God and the law. They were equal every which way. Nobody was smarter than anybody else. Nobody was stronger or quicker than anybody else. All this equality was due to the 211th, 212th, and 213th Amendments to the Constitution, and to the unceasing vigilance of the United States Handicapper General." In Vonnegut's dystopia, smart people are made to wear little receivers in their ears so that the government can send them sharp bursts of sound every 20 seconds to interrupt their thoughts, newscasters all have speech impediments (when no one can understand a news bulletin on the TV, the hero's mother comes to the announcer's defense with "That's all right . . . he tried. That's the big thing. He tried to do the best he could with what God gave him."), and beautiful ballerinas wear hideous masks and are weighted down with heavy bags of BBs.[6]

As silly as Vonnegut makes this sound, if concerns about social unrest were serious enough, it may be necessary to prevent genetically engineered people from taking advantage of their special abilities as a necessary stopgap measure until genetic engineering became affordable for everyone, since the alternative would be to physically strip them of their genetic advantages. George Annas refers to a tale by H. G. Wells: "As H. G. Wells made clear in his *Country of the Blind*," Annas relates, "it is simply not true that every enhancement of human ability will be universally praised: in the valley of the blind the one-eyed man was not king, rather eyes were considered a deformity that had to be surgically eliminated so that the sighted person would be like everyone else."[7]

Germ line evolutionary engineering for the well-off is especially corrosive to the survival of liberal democratic structures and would recreate precisely the sort of hereditary obstacles that we thought we had put behind us. Never mind that, as I explained in *The Price of Perfection*, equality of opportunity is largely a myth—as Stephen McNamee and Robert Miller Jr. point out, "the most important determinant of where people end up in the economic pecking order of society is where they started in the first place."[8] The American dream may be a fantasy, but it is an enduring fantasy. A 2007 Associated Press / Ipsos poll found that a majority of Americans agreed that "almost anyone can get rich if he puts his mind to it."[9] This delusion is fortunate, since a belief in equality of opportunity is a necessity if social stability is to be maintained in the face of so much frank inequality. By bursting this bubble, evolutionary engineering reserved for those who were already privileged could unleash a flood of pent-up envy and ignite bitter class warfare, which ultimately could destroy the conditions that sustain progressive social conditions.

In my previous books, I offered two recommendations for how to prevent the creation of a socially destabilizing genobility. In *Wondergenes*, I suggested establishing a government-run lottery. Every adult who had not already won would be entered automatically in the drawing, and winners, selected at random, would be entitled to subsidized access to whatever genetic engineering they desired. I also pointed out that the government could adjust the odds of winning as needed to maintain social stability, so that envy of the well-off who were able to engineer their children with their own money did not lead to socially destructive unrest.

In *The Price of Perfection*, which dealt with improving human performance, I argued that if safe and effective biomedical technologies were developed that significantly enhanced those human capacities that were key determinants of a person's social success, such as intelligence and charisma, it would be necessary to go beyond a lottery to maintain a sufficient amount of social cohesion. Instead, it would be necessary for the government to subsidize access to those technologies for everyone. But although a basic package of genetic interventions might be made universally available, there are bound to be other genetic interventions that only the wealthy could afford. Moreover, while universal access to germ line engineering would surely be the fairest solution and the one most likely to avoid social disruptions stemming from inequality of opportunity, it also would be expensive. As noted earlier, it now costs approximately $50,000 to produce a child using IVF, which would be a prerequisite to engineering offspring under current technological constraints, and this price does not include the cost of the engineering itself. New technologies such as nanotech could reduce costs substantially, since they may make it possible to genetically manipulate embryos that were fertilized naturally. But even at $5,000 or $10,000 a pop, subsidizing access would likely be prohibitive, given how many families would be unable to afford evolutionary engineering on their own. Therefore, a lottery that only subsidized access for some parents would be much more affordable. One factor that could drive the government to attempt a more universal approach to public access, however, would be global competition, since it is not far-fetched to imagine that some countries would decide to pay for as many people as possible to engineer their children in ways that made them more valuable to the state, and that other countries would feel compelled to follow suit to avoid falling behind in the scramble for scarce global resources.

A second way discussed in chapter 6 in which evolutionary engineering could disrupt social harmony would be if radical biological change took place so rapidly that social structures could not coevolve fast enough to accommodate them. One example would be a sudden, dramatic increase in human longevity, say, an extension of the average life span to 150 in just one generation. It is easy to envision all sorts of social disruptions ensuing, including damage to the nuclear family, breakdowns in the health care and welfare systems, and rancorous rivalries in the workforce. Requiring that adequate testing take place before professionals were free to provide evolutionary engineering to parents would retard the pace of technological advance to some extent. Beyond that, trying to slow down evolutionary engineering in the face of demands for longer, healthier, higher-quality lives, however, would be both quixotic and imprudent. The best way to cope with a jump in longevity, for instance, would be to try to anticipate how society might successfully adjust to these changes and lay in a store of readily deployable measures to facilitate the adjustments. Social scientists working with a series of different scenarios could devise a set of potential new approaches to combat age discrimination, for example, that could be implemented quickly if needed.

Chapter 6 also emphasized the importance of considering the sensibilities of people who are repelled by evolutionary engineering on religious or moral grounds. Many of these people will never accept these technologies, but they should be given as little ammunition as possible in their fight against scientific progress. Researchers trying to develop and improve engineering techniques must be careful, for example, not to violate the ethical and legal rules that have been established to govern human experimentation. Researchers also must take care to explain why a proposed step forward is not too ambitious in view of prior work. Furthermore, researchers must be mindful of the especially strong distinction the public makes between humans and other animals; most people think that, no matter how many successful experiments have been performed on other species, including primates, experimenting for the first time on human subjects crosses a moral Rubicon. Combining human and nonhuman traits would be even more unsettling. This isn't to say that such experiments should never take place, but rather that researchers must prepare the groundwork socially as well as scientifically. "The challenge facing human geneticists," said the outgoing president of the American Society of Human Genetics in 1996, "is to find the proper balance between the hopes and fears of society and the goals and interests of our science."[10]

⌒

Another method by which parental engineering decisions could undermine social cohesion would be if it diminished children's moral capacity. This goes beyond a concern that lack of sensitivity to the physical damage that genetic engineering could produce in children could lead to morally repugnant reproductive decisions, or that allowing the government to force parents to install certain traits in their children in order to make them more globally competitive would turn the children into the means to economic and political ends and start down a path toward making reproduction a public service rather than a private decision. Both of these actions would raise moral objections. The threat here, however, is far more fundamental: that evolutionary engineering would make it impossible for people to judge whether these actions were moral or immoral. As Ronald Dworkin says, "the terror many of us feel at the thought of genetic engineering is not the fear of what is wrong; it is rather a fear of losing our grip on what is wrong."[11]

Sociobiologist Edmund O. Wilson and others locate an individual's sense of values in the limbic system, a brain structure buried beneath the cerebellum, which leads Oxford philosopher Jonathan Glover to worry that genetic engineers inadvertently might make biological modifications that disrupted this system.[12] To take another example, a considerable amount of research has examined the use of certain drugs, especially the beta blocker propranolol, to treat posttraumatic stress disorder (PTSD). Immediately after an event in one's life, there is a period in which the memory of that event is stored in the brain. In the case of highly traumatic events, the body releases too much adrenaline, which produces an increase in the neurotransmitter norepinephrine. This in turn can cause a memory to be "overstored," as a result of which it vividly resurfaces, sparking the symptoms of PTSD.[13] Propranolol blocks the effect of norepinephrine, thereby preventing this overstoring. The question is whether the drug could go further and erase or prevent the storage of memories of bad events entirely. One potential use of propranolol is to fortify combat soldiers against PTSD. But critics fear that giving soldiers propranolol before they went into combat could undermine their moral inhibitions. In the words of the president's Council on Bioethics chaired by Leon Kass, they could turn into "'killing machines' (or 'dying machines'), without trembling or remorse." Preventing PTSD is a worthy cause, the council concedes, but doing it in this way would be "at the cost of making men no different from the weapons they employ."[14]

Instead of using drugs such as propranolol, evolutionary engineering might make it possible to reduce the risk of PTSD in soldiers genetically. If so, then some parents, particularly those from military families, might be tempted to modify their children in this way to give them a better chance at a successful military career. Consider the Scotts. "We've had a soldier on active duty in the U.S. Army in my family for more than 130 years," explains retired Army major general Bruce Scott.[15] His wife Mary's father was a West Point graduate who was killed in Vietnam. All six of their children are in the military, five in the Army; three attended West Point. Military service, says General Scott, "is in my family's DNA." Families like the Scotts might feel that "optimizing" their children for the military was merely carrying on the family tradition. Giving these children greater visual acuity or making them stronger, faster, or better able to survive battle damage might be acceptable, but not decreasing their moral intuition; powerful weapons should not be placed in the hands of soldiers who are morally compromised.

At least military parents would be engineering their children for what they viewed as public service. But other parents might pare down their children's moral reasoning for reasons that were hardly praiseworthy. Think of the overly competitive parents who seem willing to go to any lengths for their children, such as the notorious Wanda Holloway, who was sentenced in 1991 to a 15-year prison term for plotting to kill her daughter's chief rival in cheerleading competitions,[16] or the father of a 10-year-old hockey player who, angered that the players were being too rough on his kid, beat a coach to death in 2003. These may be extreme cases, but there are plenty of stories of parents who brutalize referees, coaches, umpires, and the children themselves. The National Association of Sports Officials claims that it receives over 100 reports of incidents such as these each year.[17] A survey of 23,000 adults in 22 countries found that 60 percent of Americans reported witnessing bad parental behavior at youth sporting events, the highest rate in the world, although India (59%), Italy (55%), Argentina (54%), Canada (53%), and Australia (50%) were right behind.[18] So it's not hard to imagine that some parents would be interested in making their children less morally inhibited in order to be more savage in sport. Still other parents might dampen their children's moral inhibitions to make them more ruthless in business or more successful at crime.

Intentionally diminishing someone's moral capacity so that they can take greater advantage of others produces no redeeming social value and should be outlawed. Since the genetic changes that would achieve this would be known, they could be screened for as part of newborn screening. If the changes were reversible, then the law should require that they be reversed; these modifications would be sufficiently destructive to society that reversing them might be justified even at a considerable health risk to the child. The health care professionals who assisted the parents should be severely sanctioned. Moreover, given that, as noted earlier, parents who engineered their children in impermissible ways ordinarily would not be penalized, since jailing the parents would unfairly deny the innocent children parental care while fining the parents would unfairly deprive the children of parental largesse, parents whose disregard for the welfare of other members of society was so extreme that they engineered their children to be immoral might well lose their right to bear additional children.

In seeking to preserve social cohesion, then, we must employ additional means beyond those necessary to protect the health and psychosocial well-being of the engineered children themselves. Parents who can afford expensive forms of evolutionary engineering do not harm the children whom they engineer, but instead the children whose parents are not as well-off and the societal institutions of which both types of children are a part. Parents who blunt their children's moral intuition may actually improve their children's welfare by making them better able to take advantage of others. If evolutionary engineering is to proceed with as little risk as possible, threats to society require their own set of regulatory responses.

CHAPTER TEN

Providing for Our Descendents

Protecting future generations from evolutionary harm not only is a challenge distinct from protecting the children who are directly engineered or the institutions of civil society but is complicated by the difficulty of predicting downstream genetic harm. The outcomes of a number of genetic experiments can look good at first but result in injury to subsequent generations. Recall that some reports say that Ananda Chakrabarty couldn't commercialize his recombinant-DNA-engineered, oil-digesting bacterium, the invention that led to the patent case in which the Supreme Court ruled that individuals could patent life-forms, because the bacterium lost its engineered ability to digest oil spills after several generations. Similarly, the benefits from genetic changes might be lost over time, but the injuries might remain, or serious injuries that were not anticipated could occur that outweighed any continuing benefits.

One solution would be to require experiments to continue over a number of generations to prove that a genetic modification was safe. As David Sloan Wilson points out, "to solve the problem of unforeseen consequences, we need to be suitably humble about what we know, cautious about implementing new technologies, and diligent about discovering unforeseen consequences. The ultimate solution to partial knowledge is complete knowledge."[1] Since, in order to be ethical, experiments in humans have to be preceded by animal experiments, one question is how long animal experiments need to last. FDA requirements for testing the safety of food additives normally only require animal studies over two generations; adding a third generation is only recommended if "overt reproductive, morphologic [structural], and/or toxic effects" show up in the second generation (the offspring of the animals originally fed the additive).[2] Sometimes longer, multigenerational studies have been conducted, however. Caffeine has been tested in as many as four generations of rats.[3] A study of genetically modified corn by the Austrian government also lasted over four generations of mice.[4] Owing to concerns about the long-term effects of ethinyl estradiol, an estrogen that is commonly found in birth control pills, the FDA's National Center for Toxicological Research required a five-generation rodent study; interestingly, the researchers saw a small increase in tumors in male sex glands in the third generation.[5] A separate question is in how many animal species tests must be performed. Multigenerational testing would be feasible in animals that live a short time, such as rodents, but what about longer-lived animals, such as primates, whose biology is more similar to our own? As Susannah Baruch and her colleagues observe, "given that it may take sixty to eighty years to obtain multigenerational data from some animal species, questions exist about whether animal data would ever be sufficient to warrant human clinical studies."[6]

As noted in the previous chapter, moreover, long-term animal experiments alone, even in primates, would not likely be deemed sufficient to provide evidence that genetic modifications in humans would not harm future offspring; long-term human experiments also would be necessary. But this would present serious challenges. For example, the FDA is grappling with how to show that drugs are safe to treat chronic ailments over a long period of time. One drug used for chronic illness is Fosamax. Recently, the FDA discovered that the class of drugs to which Fosamax belongs—bisphosphonates, which are taken for years by older women to prevent osteoporosis, the loss of bone density that causes fractures in the elderly—can eventually cause thigh fractures and jawbone degeneration, in other words, actually weaken bones rather than strengthening them. Ironically, television ads for the drug Fosamax tell viewers that "this is an amazing age. We learn something new every day."[7] In the TV ad for another one of the bisphosphonates, Boniva, Sally Field chirps that "studies show after one year of Boniva nine out of ten women had better bone density,"[8] without mentioning that the problems show up later. "The difficulty," writes Gina Kolata in the *New York Times*, "is in figuring out how to assess the safety of drugs that will be taken for decades, when the clinical trials last at most a few years."[9]

If the FDA is struggling with how to assess the safety of medical interventions over a lifetime, imagine how much more difficult it would be to discover adverse effects of genetic engineering that could afflict future generations. There would have to be some way to conduct multigenerational human investigations, that is, to evaluate the offspring of those who had been modified. Since these descendants in effect would be experimental subjects, presumably they or someone with the proper authority would have to give permission for them to be studied. Parents who participated in engineering experiments could agree to allow the resulting children to be examined and monitored, but parental permission would lapse once the child reached adulthood. As Baruch et al. point out, "researchers may want prospective parents to agree to have their children, and perhaps several generations thereafter, studied from birth. However, it would not be possible to guarantee participation in a study, as participants are always free to withdraw."[10] So there would need to be some way to keep track of the children so that they could be asked to give their permission when they became adults.

The problem of staying in touch with descendants has arisen in a more limited fashion in connection with genetic testing for disease risks. In a couple of court cases, children have sued physicians claiming that they should have been given information about their parents' genetic illnesses so that they could have taken steps to protect themselves. In one case, a Florida physician group that had treated a patient for a hereditary form of cancer was sued by the patient's daughter when she contracted the same cancer three years later. The daughter claimed that the physicians should have warned her that she could be at risk, which would have enabled her to take preventive measures. The Florida Supreme Court rejected the doctors' argument that they owed their duty only to their patient and not to the patient's daughter and held that the doctors should have advised the mother to inform her daughter.[11] A second case in New Jersey involved the daughter of a patient who died of colon cancer. The daughter, who contracted the same cancer, claimed that her father's doctor should have told her that his cancer was heritable, in which case she would have had her colon removed and avoided her father's fate. The New Jersey appellate court agreed and went further than the Florida court by suggesting that the father's doctor might have been obligated to warn the daughter directly, rather than just advising her father to warn her. What was especially bizarre about the New Jersey case was that the father had been diagnosed originally in 1956, when his daughter was 2 years old, and had died when she was 10. The daughter did not discover that she had the same disease as her father until she was 36, and she filed her lawsuit two years later. By that time, her father's doctor, like her father, had died, and the daughter actually ended up suing the doctor's estate.[12]

Keeping track of what happens to genetically modified children and their offspring presents legal as well as practical problems. Descendants would have to be informed that their parents had been engineered, but this could breach patient-physician confidentiality by revealing personal information about the parents.[13] That was one of the reasons mentioned by the court in the Florida case for why it would only require physicians to advise their patients of the need to tell their children, rather than requiring the physicians to inform the children directly as the New Jersey court suggested.

Assuming that some registry was established or some other way devised to trace the offspring of engineered parents, how many generations would need to be followed before a genetic alteration was deemed safe? Would it be the same as for animal studies—usually two or three generations? Would the number of generations change depending on the type of modification, becoming longer for alterations that were thought to pose greater downstream risks? The problem is that the more generations that needed to be followed, the more difficult it would be to keep track of descendants. I am reminded of a cartoon in which one researcher holds a flask up to another and says, "It may very well bring about immortality, but it will take forever to test it."

᠕

Transhumanists are well aware that requiring long-term longitudinal studies in humans could make it much harder for evolutionary engineering to become accepted. The World Transhumanist Association therefore cautions that "the standards for the safety of inheritable genetic therapy should be no higher than the safety of unassisted reproduction. Multi-generational effects may be explored in animal trials, but human trials should not have to demonstrate a low risk of teratogenic effects [birth defects] beyond the first generation." The association also resists the creation of a mandatory registry of altered individuals.[14] Conceivably testing methods could be developed in the future that would obviate the need for long-term human studies. In the 1970s, in response to the need for a rapid way to determine if a chemical such as a food additive posed a cancer risk to humans, a group of Berkeley researchers led by Bruce Ames developed a bacterial test that they claimed could accurately detect carcinogens without the need for experiments in rats and mice, which led to the test being embraced by animal rights advocates. But Ames's group was able to demonstrate the accuracy of its bacterial assay by comparing it with results from numerous rodent experiments that already had been conducted. There would be no similar benchmark for an alternative test for human evolutionary engineering unless a substantial amount of long-term human investigation previously had taken place.

In the face of these obstacles, transhumanists and parents eager to engineer their children may lobby for relatively weak regulatory regimes in which little multigenerational experience in humans is required before the use of an engineering technique is allowed to become widespread. A more relaxed approach may be appropriate for germ line manipulations aimed at preventing, treating, or curing serious genetic diseases as opposed to minor ailments and nondisease characteristics, since there would be greater urgency and more potential benefit to offset incomplete safety information. A historical analogy is the campaign by AIDS activists to force the FDA to relax its rigid stance against allowing patients to receive experimental drugs that had not completed clinical (i.e., human) testing. Johns Hopkins health law researcher Gail Javitt describes this campaign as "an unprecedented demonstration of grassroots protest against the agency, [in which] FDA's Rockville, Maryland headquarters were picketed by protestors, a thousand strong and many suffering from AIDS, who chanted for hours demanding faster drug approval. Protesters also took more subtle, and potentially more damaging, measures to demonstrate their unwillingness to accept the established clinical trial system as a means to drug approval."[15] Javitt recalls that one AIDS activist group figured out a way to analyze the ingredients in the pills given to patients in clinical studies so that the patients could tell whether they were getting the experimental drugs or a placebo. This effectively halted the experiment, since it made it impossible for it to remain double-blinded, that is, to maintain the critical condition for valid assessment that neither the experimenters nor the subjects know what it is that specific subjects are being given.

Activists took other steps to frustrate the experiments, Javitt explains: "Patients receiving active treatment shared their drugs with those receiving placebos. Patients also adjusted their doses and added treatments prohibited by the protocol. Finally, an 'underground' distribution network developed for drugs being tested in clinical trials."[16] The AIDS community's campaign achieved substantial success. The initial result was an effort by the FDA to speed approval of AIDS drugs, which culminated in a 1987 regulation permitting drugs undergoing but not yet finished with testing to be distributed to patients with severe illnesses. This was followed by a 1988 rule in which the FDA declared that it was willing to relax its traditional requirement that drugs must complete all three phases of clinical testing before they could be marketed; henceforth, the agency announced that it would allow certain drugs to be marketed after they had completed only the first two phases, foregoing the most expensive and time-consuming Phase III trials in large patient populations. The FDA adopted other measures to speed up drug approvals in the ensuing years, and in 1997, Congress amended the Federal Food, Drug, and Cosmetic Act to incorporate many of the agency's approaches. Therefore, it is easy to imagine an alliance of committed transhumanists and parents whose children are at risk for genetic disorders resorting to similar political tactics to get the government to back off insisting on stringent testing requirements, at least for evolutionary engineering aimed at preventing illness.

Still, having at least some long-term data about the effects of germ line modifications in humans may be imperative even when they target genetic ailments. There would be little point in removing one serious disease from an evolutionary line only to substitute another. Therefore, absent some extremely compelling reason, such as a virulent pandemic or a radical environmental shift that demanded rapid adaptation to avoid a serious threat of extinction, lax safety standards that did not require an appropriate amount of longitudinal data must be resisted, and means should be found to circumvent the practical and legal hurdles that long-term studies present. Confidentiality seems the most easily surmountable obstacle. Systems have been proposed to track children who are produced with the aid of donor eggs or sperm so that they can be informed about genetic risks that they may have inherited and so that other children conceived with reproductive cells from the same donor can be alerted to any latent health hazards.[17] Parents who wished to participate in genetic engineering experiments could be required to agree to keep the investigators aware of their children's whereabouts until the children reached adulthood, at which point the investigators could ask the children themselves for permission to continue to monitor them and their offspring, who in turn would be asked for permission to be followed, and so on. Some descendants no doubt would refuse, but long-term surveillance would have to continue until enough affected individuals had agreed to be followed that adequate multigenerational safety data had been accumulated.

In 1991, Marc Lappé, a toxicologist who campaigned vigorously against genetically modified foods and other forms of evolutionary engineering, did offer one draconian alternative to multigenerational safety data. "It is also theoretically possible," he said, "to limit the effects of germ line manipulations to a single generation, either through concurrent manipulations that limit fertility or by committing the conceptus to abstain from reproduction as one of the trade offs of his genetic alteration."[18] In other words, genetically engineered children could be deliberately rendered sterile, or be required to promise not to have children. Lappé notes that these are only theoretical possibilities, however, since designing children to be sterile would be ethically and politically unacceptable, while there would be no way to enforce a promise to refrain from having children in the future short of sterilizing genetically engineered persons before they reached child-bearing age or forcing them to abort, neither of which can be justified.

CHAPTER ELEVEN

Safeguarding the Human Species

In addition to protecting specific future persons from harm caused by the genetic modification of their forbearers, it is obviously also imperative to avoid imperiling the human lineage and its cumulative genetic inheritance. The protections against harming children will considerably reduce the risk of this happening by helping to prevent parents' use of evolutionary engineering techniques that have not been shown to be safe and efficacious, especially if multigenerational studies are required. But could the survival of the lineage be threatened by genetic modifications that did not cause harm to children or specific descendants? In other words, would the measures that prevented harm to children also adequately protect the lineage?

I have addressed several ways in which evolutionary engineering could threaten the continuation of the lineage: loss of genetic diversity due to large numbers of parents making the same modifications in their children's genes, excessively large body sizes, inability to reproduce, and gender imbalance. Would these be proscribed as being harmful to children? Making children gigantic is likely to cause them emotional if not physical harm as well. Limiting their ability to reproduce also seems like it would be injurious to children. But what about selecting the child's gender? Deliberately having more male than female children could create a risk for the lineage, but it is doubtful that anyone would consider being born a male to be harmful to the child, let alone deleterious enough that it warranted interfering with parental decision making. In societies with strong preferences for males, being designed *female* might raise eyebrows, but being male would seem to be advantageous rather than detrimental. Therefore, there would be no call to prevent parents in Asia or elsewhere from engineering more males than females, or even all males, on the basis that it caused harm to the child. So even under a rule that prohibited evolutionary engineering that harmed children who were engineered or their specific descendants, an extreme gender imbalance might arise that threatened the survival of the lineage.

Another threat to the lineage that would not seem to count as harmful to engineered children would be if they were modified in such a way that they could only mate successfully with children who also had been engineered, which might be very attractive to parents who wanted to create or preserve a genobility, but which could lead to a societal breakdown by igniting conflict between classes, subspecies, or separate species. Moreover, although a herd mentality among parents in terms of how they designed future children might jeopardize genetic diversity by producing too many children with too many of the same genetic modifications, it is difficult to construe being made to be like everyone else as a potential cause of harm to children.

So even if children and their offspring were protected against evolutionary engineering that might harm them, some serious threats to the lineage could slip by. One noteworthy loophole in the regulatory regime that protects human subjects, for example, is a provision in the Common Rule that could be interpreted to bar IRBs from considering adverse effects on the human lineage as an experimental risk to be weighed against the potential benefits. Specifically, the rule says that "the IRB should not consider possible long-range effects of applying knowledge gained in the research (for example, the possible effects of the research on public policy) as among those research risks that fall within the purview of its responsibility."[1]

This is a highly perplexing provision in the federal regulations, and to find out why it was included, I contacted Bradford Gray, a health policy expert at the Urban Institute in Washington, D.C., who was on the staff of the National Commission for the Protection of Human Subjects of Biomedical and Behavioral Research, the body that drafted the Belmont Report that was the source of the Common Rule. According to Gray, the prohibition against considering long-range risks stemmed from a number of concerns. Reminiscent of Cass Sunstein's position that using the precautionary principle to decide whether or not to go forward with a research project that poses unknown or uncertain hazards would be overly cautious, the commission felt that long-range risks were too speculative. It is always possible to imagine dire consequences that could flow from an experiment, explained Gray, yet it is impossible to know how much weight to give them. The commission also wanted IRBs to focus on protecting the actual experimental subjects, rather than on controlling what research ought to take place by trying to anticipate the net long-term benefits from an experiment. Another concern was that

IRBs did not have the expertise to undertake these sorts of technology assessments, while there were numerous other entities such as the NIH that had both the expertise and the responsibility. The commission also recognized that, as borne out by the doomsday scenarios surrounding particle accelerators and other research enterprises discussed in chapter 2, long-range harms from scientific research extended well beyond research on human subjects, which is the focus of the IRBs. But according to Gray, the major concern that led the commission staff to include the restriction on the scope of IRB review was the desire to prevent controversial social science research such as research on race and intelligence from being blocked on the grounds that it might have long-term, adverse societal effects (Bradford Gray, pers. comm., Mar. 19, 2007).

Assuming that the type of concerns that led to this provision in the IRB regulations could be allayed or ignored, the Common Rule could simply be rewritten to authorize IRBs to consider threats that evolutionary engineering experiments posed to the survival of the lineage. But a good deal of evolutionary engineering would take place outside the experimental realm and therefore free from IRB oversight. As noted earlier, under current law, once the FDA approves a new technology, doctors are free to use it for any purpose they wish. So unless the law was changed in this respect, an experimental modification that did not pose a danger to the lineage might pose such a danger when used in another way after it had been approved. And since the modification would not present a sufficient threat of harm to a specific individual, protections for current or downstream children would do little to stop it from taking place.

How then can we prevent parents from engineering their children in such a way that a severe gender imbalance or loss of diversity resulted or the children were unable to mate with humans who had not been engineered? One answer would be to rely on health care professionals to regulate themselves. As noted earlier, do-it-yourself germ line genetic engineering seems unlikely, and therefore parents will have to obtain the services of professionals such as physicians and geneticists in order to accomplish their engineering objectives. We also noted that professional groups such as the AMA have begun to establish guidelines to govern genetic engineering performed by their members. So far the guidelines have not addressed concerns about saving the human lineage, but there is no reason why they couldn't be expanded to take this into account along with the health and well-being of the children. But professional self-regulation is not sufficiently dependable to be relied upon exclusively. Biotech researchers will be eager to achieve momentous breakthroughs, such as Edwards and Steptoe's development of IVF that led to Louise Brown's birth in 1978, for which Edwards won the 2010 Nobel Prize in Medicine (Steptoe having died in 1988); Ian Wilmut's announcement in 1997 that he had successfully cloned Dolly the sheep from one of its mother's mammary cells; or Craig Venter's 2010 claim, albeit somewhat exaggerated, that he had created the world's first synthetic life form, "a defining moment in biology" that "heralds the dawn of a new era in which new life is made to benefit humanity."[2]

The allure of fame and fortune that accompanies these advances will tempt professionals to ignore ethical and legal norms. Martin Cline, a researcher and chief of Hematology/Oncology at the UCLA Medical Center, flew to Italy and Israel in 1980 to conduct gene therapy experiments before they were approved by the UCLA IRB, although he was eventually stripped of his UCLA department chair and disqualified from receiving future research funding from the NIH.[3] Although human cloning was generally deemed to be impermissible, physicist Richard Seed announced in 1998 that he would clone his third wife at a new facility he was building in Hokkaido, Japan,[4] and fertility doctors Panayiotos Zavos and Severino Antinori declared that they would attempt to clone a human being and that, if denied permission, they would conduct their experiment on a boat in international waters.[5] Professionals also will be attracted by the money to be made by assisting parents in designing their offspring. Manufacturers of genetic engineering equipment and supplies will mount marketing campaigns similar to the ones now conducted by drug companies, which employ one sales representative for every three physicians in the country.[6]

If professional self-regulation cannot be trusted to completely control evolutionary engineering on its own, then governmental oversight of evolutionary engineering would have to be extended beyond protecting children to protecting the lineage. Jonathan Glover acknowledges that "it could be that some centralized decision for genetic change was the only way of securing a huge benefit or avoiding a great catastrophe."[7] Glover, for example, describes what he calls a "mixed system" in which parents choose their children's characteristics but the government has a veto over those that are deemed to be "deleterious." In order to preserve genetic diversity, for example, "deleterious" could be interpreted to include threats to the lineage, and parental choice could be restricted so that they did not all make the same genetic modifications. Mark Frankel and Audrey Chapman offer another option: only parents who, by virtue of having two copies of a disease gene, actually have a genetic disease should be allowed to delete genes for recessive diseases such as cystic fibrosis, sickle cell anemia, and Tay-Sachs disease.[8] That way, presumably, the genetic diversity represented by the recessive gene would persist because the gene would be inherited from parents who possessed only one copy and therefore were "carriers" for the disorder rather than actually afflicted with it. Similarly, speciation could be prevented by prohibiting parents from modifying children in such a way that they could only mate with other designer children. Sex selection could be limited or banned to avoid an imbalance of genders that could push humans past an evolutionary tipping point.

The question is how far government control over parental choices should go. One risk is that governments might be tempted to seize complete control over parental decision making. Earlier we discussed how governments might be interested in having parents turn out the type of offspring that would be most useful to the state, such as more productive workers or better soldiers, and it is possible to imagine some devious governments seeking to justify an extreme amount of government direction on the basis that it was necessary to save the human lineage. Or perhaps governments would start out by preventing parental decisions that were potentially lineage lopping and then begin to slip in an increasing number of affirmative design obligations. Such a degree of government interference with reproductive decision making would, to say the least, be unprecedented. It would go far beyond the Chinese government, which currently prohibits people with genetic diseases from having children, and it would even trump the Nazi racial program, which dictated who could reproduce and with whom, but lacked the scientific means to determine what specific characteristics a child would inherit. It is unlikely that people in liberal democracies would accept this amount of government intrusiveness in their reproductive behavior, and in the United States at least, unless the government was responding to an immediate, dire threat to society, it would likely be found unconstitutional.

In the previous chapter, we saw that abuse and neglect laws can form the starting point for laws protecting children from dangerous evolutionary engineering. Similarly, there is a legal point of departure for regulating engineering in the interest of the lineage. This is the body of policies, principles, statutes, and administrative regulations called public health

law. The Constitution gives the government the "police power" to protect the public from being harmed by its members, and while the police
power may be most closely associated with law enforcement, one of its
most important responsibilities is to protect the public's health. It would
seem like the destruction of the human lineage would be a consummate
threat to public health. To corrupt an old saying about the importance of
health, if you don't have your lineage, you don't have anything.

Public health authorities currently have very broad powers. They can
sequester not only people who are known or suspected of having a transmissible disease but those who merely have been exposed to the disease,
such as by having traveled in a country where it is found. People incarcerated in this way, called "quarantine," can be held for as long as necessary
to ensure that they get over the disease and are no longer contagious,
until they can demonstrate that they were not infected in the first place,
or, as in the case of Mary Mallon—aka "Typhoid Mary," who spent a total
of 26 years locked up on an island in the East River—until they die. In the
1990s, New York City confined for approximately six months over 200
people who refused to be treated for drug-resistant tuberculosis (TB).[9] In
addition to quarantine, public health officers can invade people's privacy
by requiring them to reveal the identity of those with whom they have
come into contact, a practice known as contact tracing. Contact tracing
has been used in an attempt to combat the spread of HIV, for example,
particularly in San Francisco; the contacts in that case were sexual partners, illustrating the degree to which individual privacy is compromised
in the interest of protecting the public's health. In addition to quarantine
and contact tracing, public health officers can forcibly treat people, com-
pel them to be vaccinated, and obtain a sample of blood from a newborn
before it is allowed to go home with its parents. Added to all this is the
power of the states to pass laws defining unhealthy behaviors, which can
run the gamut from operating an unsanitary restaurant kitchen to def-
ecating in public, and to punish violators. A current public health target
is obesity, with fast food as a prime villain; Los Angeles banned fast food
restaurants in areas of the city with high rates of obesity such as South
Los Angeles, while Santa Clara County prohibited fast food restaurants
from selling meals with toys.[10]

The government's public health power would seem to be more than
adequate to enable it to prevent parents and the health professionals
from whom they obtain the necessary technical wherewithal from engineering children in such a way as to jeopardize the continuation of

the human lineage. The government could pass a law making it illegal to modify children in ways that threaten the lineage, bestow on public health officials the authority to inspect the records of clinics where evolutionary engineering takes place, use mandatory newborn screening and other investigative methods to track down offenders by identifying children who had been illegally engineered, and punish the professionals who violate the law.

Relying on the government's current public health powers to save the lineage would encounter some obstacles, however. First, public health concerns tend to focus on relatively immediate threats, such as epidemics, food-borne illnesses, and sexually transmitted diseases. There is no precedent for using public health powers to protect against risks to future populations, much less highly speculative risks to remote descendants. True, sanitation systems are supposedly built to last a while; in 2003, Los Angeles replaced a water delivery system that was almost 100 years old.[11] But upgrading sanitary infrastructure also provides immediate benefits to current residents. It is unclear how long-range the goals of public health officials could be and still have courts uphold coercive state action as within the current reach of the government's police powers.

A second problem with relying on the public health system to protect the human lineage is that its powers historically have been vested in the states rather than in the federal government. Relying on states to adopt protections for the lineage could lead to a patchwork of rules around the country that could hinder enforcement, and some states could enact more lenient laws in order to attract or retain the evolutionary engineering industry, similar to how California, New York, and Massachusetts welcomed and subsidized research on embryonic stem cells when President Bush banned federal funding. To date, the main public health focuses of the federal government have been to produce and disseminate information, the central mission of the Centers for Disease Control and Prevention (CDC),[12] and to protect the nation from health threats from abroad, such as by preventing persons with communicable diseases from entering the country. Nevertheless, the federal government also has considerable authority to control interstate commerce and has used this power in furtherance of the public's health, for example, by regulating foods and drugs. There is some disagreement over whether this federal authority is as extensive as the states', with some commentators maintaining that the federal government may only regulate public health matters that involve economic activity, like the marketing of foods and drugs or the sale of marijuana for medicinal use.[13] If they are correct, then the question is whether protecting the fate of the human lineage can be considered "economic" regulation. Since the end of the lineage would be the end of everything human, including commerce, the answer might seem to be "yes," unless someone wanted to make the somewhat bizarre argument that commerce could still take place between the members of other lineages. In any event, the federal government clearly has the authority to give states money on certain conditions; this is how the federal government gets states to conform to federal rules for Medicaid and other federally subsidized assistance programs. So under the exercise of its spending power, Congress could offer funds to the states to police evolutionary engineering if they agreed to do so in conformity with federal guidelines.

There is a far more serious problem, however, with relying on the public health power of government to protect the human lineage from unacceptable forms of evolutionary engineering: the government has not always wielded these expansive powers judiciously. The history of the U.S. public health system is a story of great accomplishment, including the construction in the nineteenth century of urban water and sanitation systems to protect the public against filth and the mass inoculation programs in the late nineteenth and twentieth centuries that culminated in the triumph over polio in the early 1950s. But the public health system also has been the culprit in a number of the greatest injustices perpetrated by American medicine.

Given the broad powers wielded by public health officials, it is perhaps surprising that the entire legal underpinning of the public health system rests on a single 1918 Supreme Court decision. Cambridge, Massachusetts, wanted to vaccinate its residents against smallpox, but one of them, Henning Jacobson, sued the state public health department after he was fined five dollars for refusing to be vaccinated against smallpox and then was jailed when he refused to pay the fine. The court's opinion, written by Justice Harlan, is a sweeping endorsement of the government's public health powers, which Harlan analogized to the power to defend the nation against foreign attack. If the government could require its citizens to take up arms and risk "the chance of being shot down," Harlan reasoned, then surely it can require them to be vaccinated against a deadly disease. The common good, Harlan emphasized, takes precedence over the "wishes or convenience of the few," and the only limits on the exercise of these broad powers are that they may not be either "arbitrary or unreasonable" or "cruel and inhuman." While Harlan acknowledged that the state could not force someone to be vaccinated if it would "seriously impair his health, or probably cause his death," Henning Jacobson's objection that he had had an adverse reaction to vaccination as a child was not convincing enough to justify an exception in his case.

The *Jacobson* case not only laid the foundation for all subsequent public health law but also was the only precedent that the Supreme Court cited nine years later in *Buck v. Bell*, the case mentioned earlier in which the justices, with only one dissent, upheld the involuntary sterilization laws that were the lynchpin of the American eugenics movement. A woman named Carry Buck, who had been institutionalized in the Virginia State Colony for Epileptics and Feeble Minded, where she had been sterilized under the provisions of the Virginia eugenics law, filed a lawsuit supposedly to challenge the constitutionality of the statute. As legal historian Paul Lombardo discovered, however, the suit was a sham.[14] Buck was given virtually no effective legal representation; her lawyer, as well as her ostensible supporting witnesses, had concocted the lawsuit together with the state officials in order to give the courts an opportunity to approve the constitutionality of the Virginia law, which was intended to serve as a model for sterilization laws in other states.

When Buck's case reached the U.S. Supreme Court in 1927, Oliver Wendell Holmes, one of the most respected jurists in American history, upheld the law with the following, now-infamous words:

> We have seen more than once that the public welfare may call upon the best citizens for their lives. It would be strange if it could not call upon those who already sap the strength of the State for these lesser sacrifices, often not felt to be such by those concerned, in order to prevent our being swamped with incompetence. It is better for all the world, if instead of waiting to execute degenerate offspring for crime, or to let them starve for their imbecility, society can prevent those who are manifestly unfit from continuing their kind. The principle that sustains compulsory vaccination is broad enough to cover cutting the Fallopian tubes. Jacobson v. Massachusetts, 197 U.S. 11. Three generations of imbeciles are enough.[15]

Note the reference to the sole supporting precedent, the *Jacobson* case. Decades later, Lombardo's research established that Buck had not been institutionalized because she was mentally challenged, but because she had become pregnant after she had been raped by the nephew of the foster family she had been living with. Government records show that neither she nor her mother or child was in fact "feeble-minded."

◦⸻

Although the eugenics movement stands out as one of the most appalling examples of overreaching public health policy, the public health system is guilty of many other sins. It conducted the barbaric Tuskegee experiment described in the previous chapter, which was both unethical in leaving sick persons untreated and racist in being conducted on African American men. In the late 1940s, the U.S. Public Health Service also intentionally infected Guatemalan men with syphilis, ostensibly to determine the efficacy of antibiotic treatment.[16] These syphilis experiments were not the first time public health officials had taken action against sexually transmitted diseases, however. During World War I, over 20,000 women believed to be at risk for spreading syphilis, and therefore given the name "spreaders," were incarcerated in government camps, a move that Harvard historian Allen Brandt called "the most concerted attack on civil liberties in the name of public health in American history."[17] By the end of World War II, all states required syphilis testing before a couple could obtain a marriage license. Most states repealed these laws in the 1980s, but not before many people suffered severely as a result of the inaccuracy of the test, which 25 percent of the time incorrectly reported that people had the disease.

During the early phase of another sexually transmitted disease epidemic, AIDS, televangelist Jerry Falwell called for all prostitutes to be placed in quarantine, and William F. Buckley and the then vice president George Bush called for universal HIV screening. Imposing mandatory screening was thwarted by the discovery that a person could be infected with the HIV virus and yet have a negative test result because of a delay in the ability of the test to detect antibodies to the virus in the blood. This led to an emphasis instead on "universal precautions" such as surgical gloves and masks given to health care workers and the condoms that promiscuous sexually active people were told to use on the premise that you could never know whether a partner or patient was infected. The stigma attached to AIDS and the discrimination faced by people who were infected or at risk also persuaded the authorities that more people would obtain HIV testing if the tests were available on a voluntary, anonymous basis than if they were obligatory and the results required to be reported to public health officials. Nevertheless, Illinois began requiring HIV testing for marriage licenses in 1987, and although by the end of 1988 the state had screened 159,000 people at a cost of $5.6 million, only 23 cases had been detected. Moreover, many members of the public health community chafed at anonymous testing because they felt that it placed the public at an unnecessary risk by interfering with efforts to keep tabs on persons who were infected and trace their sexual contacts. There were repeated calls to halt anonymous testing, and as treatments for HIV began to be developed, one state after another shut down its anonymous testing program.

What was ironic about the call for universal, mandatory HIV testing during the mid-1980s was that such a system had been tried 10 years earlier to combat another disease, sickle cell anemia, and it had been an abysmal failure. When a rapid, accurate test was discovered for the genetic mutation that causes sickle cell disease, a number of states passed laws that required the entire population to be screened for the mutation; however, a number of states limited the screening to African Americans, who compose almost all of those who carry the disease mutation. In some states, testing was a prerequisite for children entering public school. Sickle cell disease is not transmissible through casual sex or coughing, like HIV or smallpox. Rather, it is inherited from one's parents. The idea behind the screening program therefore was that people who found out that they had the sickle cell trait could avoid having children with other people who had the trait, and in this way, the disease would eventually be eradicated. This approach could succeed because sickle cell disease is a recessive genetic disorder, meaning that a person who actually has the disease has inherited two copies of the disease gene, one from each parent. On the other hand, "carriers," that is, people with only one copy of the disease gene, don't have the actual disease, but they can pass the gene on to their children, and if they conceive the child with another person who has the gene, there is a 25 percent chance that the child will inherit one copy from each parent and therefore manifest the disease. (A similar program that relies on stopping carriers of a genetic disease from reproducing, the Orthodox Jewish campaign against Tay-Sachs mentioned earlier, has been largely successful.)

The problem with the sickle cell screening program was that it was not accompanied by adequate public education. Few people understood what it meant for a genetic disorder to be recessive, with the result that many persons who only had one copy of the gene mistakenly interpreted a positive test result to mean that they actually had the disease. Moreover, many persons who were tested did not understand that sickle cell is a disease with a "highly variable expression," meaning that the severity of the disease varies substantially from one individual to another, and that some people who are affected will only display mild symptoms; as a consequence, many of those who found out from the test that they had the disease before they noticed symptoms erroneously assumed that they were bound to get the disease in its full-blown form, which is marked by episodes of intense pain, serious infections, and organ damage. Eventually most of the mandatory screening laws were repealed, but not before many people suffered unnecessary emotional distress. And in a bizarre footnote, lest you think that winning a Nobel Prize for science equips you to make good public health policy, Nobel laureate Linus Pauling, who did pioneering work in the late 1940s on the molecular basis for sickle cell disease, urged that all carriers of the sickle cell trait be tattooed with an "S" on their foreheads so that they could avoid reproducing with another carrier.[18]

Even when it comes to dealing with straightforward public health risks such as highly communicable diseases, public health officials can overstep their bounds. In 2007, an Atlanta lawyer named Andrew Speaker reportedly ignored the fact that he had contracted a highly contagious, drug-resistant form of TB and cavalierly flew to Europe to get married. The press accounts described how the doughty CDC tracked the miscreant down in Italy and asked him to stay put while they decided on the proper course of action; Speaker allegedly did not want to get stuck in an Italian hospital, so he flew to Prague and then on to Montreal, rented a car, and drove to New York, having dodged the fact that his name was on the government no-fly list by not flying into the United States, and having lucked out crossing from Canada to the States because the border patrol ignored an order to detain him. The CDC, the story goes, finally nabbed Speaker when he eventually showed up at a hospital in New York.[19]

Speaker tells a somewhat different story, however. According to his account, he contracted TB when he visited Vietnam in 2006 as a goodwill ambassador for the Rotary Club, and while he was receiving treatment, he made plans to get married in Greece and go on a two-week European honeymoon. Two weeks before he and his fiancée were due to leave, Speaker's doctors told him that he was suffering from a rare, highly drug-resistant form of the disease, that he needed to get treated for it at a Denver hospital that specialized in cases like his, and that it would take about three weeks for them to make the arrangements. His doctors, however, told him that he would not be quarantined in Denver, because he was neither contagious nor a threat to anyone's health, and that until then he could go about his daily life. Although his doctors advised him not to go to Europe, Speaker figured that he could squeeze his wedding and honeymoon in during the time the arrangements were being made in Denver, so he went ahead with the trip. While he was in Europe, the CDC contacted him and told him for the first time that his TB was "extensively drug resistant," the most dangerous form of the disease. Later testing showed that the CDC had been wrong; Speaker in fact was infected with a less drug-resistant version.[20] Speaker eventually got to Denver and received the treatment he needed.

Speaker was accused of ignoring his doctors' advice, but he was a model citizen compared with Robert Daniels, who was diagnosed with extensively drug-resistant TB around the same time. Daniels was born in Russia, moved to Phoenix with his parents when he was a child, went back to Russia in the late 1990s, and contracted TB, probably while he was serving a jail sentence for possession of marijuana. "You could catch it anywhere," Daniels explained. "I just had a low immune system because I was, you know, partying a lot. I was young—too much beer, vodka, women, smoking."[21] At first, Daniels's TB didn't seem to be that serious. When it got worse, however, he flew back to Phoenix. After working as a manual laborer while living in his used car, he ended up in a TB residential treatment facility for homeless persons. But he failed to take the drugs he was supposed to and, as a result, developed the same extensively drug-resistant form of TB that the CDC had at first erroneously thought afflicted Andrew Speaker. Not only did Daniels not comply with his treatment regimen, but he refused to wear a face mask in public.

Daniels, like Speaker, eventually got treatment in Denver, and given his reckless behavior, it probably made sense to quarantine him until the treatment could be administered. But in Phoenix, Daniels had the misfortune to run into Maricopa County sheriff Joseph Arpaio, who describes himself as "America's toughest sheriff."[22] Despite an Arizona law requiring that quarantined persons be kept in the "least restrictive environment,"[23] Arpaio locked Daniels in the county hospital in a solitary confinement cell reserved for convicted criminals who need medical treatment. According to the American Civil Liberties Union (ACLU), which later filed a suit on Daniels's behalf, he was not allowed to go outside, look out of a window, exercise, have access to TV or the Internet, or receive visitors. He was watched by a video camera 24 hours a day, and the light in his cell was never turned off. He had no fresh air, lost 25 pounds, and was not allowed to take a shower for nine months. His ordeal only ended because of the ACLU's intervention.[24] The point is that Daniels was not a criminal, but a patient under public health quarantine. Considering Justice Harlan's statement in the *Jacobson* case that the exercise of public health powers may not be "arbitrary or unreasonable," Daniels's treatment is hardly consistent with Harlan's admonition.

The greater the perceived threat to the public's health, the more the public health system seems inclined to overreact. The latest example is the response to 9/11 and the subsequent anthrax attacks by mail. Investigations after these incidents showed the woeful inadequacies in the public health infrastructure—the network of health professionals, government agencies, hospitals and other community health care institutions, training facilities, laboratories, supplies and equipment, and information systems counted on to respond to public health emergencies. The investigations prompted an immediate, large-scale increase in public health spending. But additionally, many public health professionals felt that they did not have adequate legal authority and discretion to respond effectively to bioterrorism. So they obtained funding from the Robert Wood Johnson Foundation and the W. K. Kellogg Foundation to support the drafting of a new model law for states to adopt. The initial version of the law, however, was vastly overreaching, and the drafters had to scale it back considerably in response to blistering criticism from human rights advocates. Just to take one example, under the original language, state or local public health officials, acting entirely on their own initiative, could have imposed compulsory screening for any health condition that they regarded as a serious threat to the community.[25] This would have allowed them, for instance, to require all pregnant women to undergo prenatal genetic testing to determine the health status of their fetuses, a step that appears to have been endorsed by the American College of Obstetrics and Gynecology for conditions such as Down's syndrome, but which has been steadfastly opposed by pro-life and disability rights groups. The original version of the law also would have authorized public health officials to cut off Medicaid and welfare benefits for pregnant women who refused to be tested.[26]

So while the powers vested in the public health system seem expansive enough to protect the human lineage, the risk is that they would be applied arbitrarily or unreasonably, as Justice Harlan cautioned. So how can the government use its public health powers to protect the lineage without overstepping its bounds? There are a couple of options. One is to employ a so-called command-and-control approach similar to the one recommended in the previous chapter for dealing with engineering that harms children. Reasonable laws could be enacted that prohibited parents from selecting their children's sex, deliberately producing children who could only mate with other engineered children, or making too many of the same engineering choices as other parents. These laws could be enacted proactively, or public health officials could establish tolerances for gender ratios and common engineering choices and only impose restrictions if the tolerances were exceeded. Children's DNA could be sequenced as part of newborn screening. Once illicit acts were discovered, the authorities could order the parents to identify the health professionals who performed the engineering, who would then face sanctions.

An alternative to a command-and-control approach would be for public health authorities to adopt an incentive program in which parents who conformed to government reproductive guidelines would receive tax credits or other economic benefits. This would not control parental engineering behavior completely, since parents who felt strongly enough simply could forego the incentives and proceed as they wished, but it might reduce the number of troublesome births sufficiently that they dropped below the threshold necessary to imperil the lineage. If the incentive approach appeared to be failing, more coercive measures could be implemented.

But let's look more closely at some of the practical problems. It might be possible for public health officers to use genetic testing or physical examination to ascertain that parents had made their children capable of mating only with other engineered individuals. But how would the public health authorities know that parents had engaged in sex selection? A requirement could be imposed on health care professionals to report the actions they had taken at parents' behests, but a professional who was willing to assist parents in engaging in unlawful sex selection would be unlikely to admit it. Public health officials could discern from birth records whether a gender imbalance was occurring at the population level, but how would they determine which parents were responsible for it, that is, which parents had made deliberate choices about the sex of their children? Determining whether too much genetic diversity was being lost because parents were making too many of the same engineering decisions might be somewhat easier, since newborn DNA could be sequenced during newborn screening and compared with the DNA of other newborns. But avoiding too much uniformity would seem to require the government to keep score of the engineering decisions that parents and professionals were making and getting enough parents to make different decisions so that a sufficient amount of genetic diversity was maintained. Of course, to do this, it would have to be possible to calculate how much genetic variation would be sufficient to assure the survival of the lineage, which in turn would require anticipating what future environmental threats the lineage might face and how much diversity would be necessary to enable humanity to overcome them.

At the present, answers to these questions would be highly speculative, although they might become less uncertain as our knowledge increases. But even assuming that we figured out how much diversity was needed, public health officials would have to be able to make parents make sufficiently diverse engineering decisions, and this might only be possible if parents were required to obtain a permit before they modified a child. The parents then could be given a certain number of selections to make from a government list, and the lists could be changed for different parents based on the results of newborn screening so that parents were making modifications that preserved a sufficient amount of genetic diversity. The permit also could instruct parents what gender to select if they wished to choose their child's gender.

Would public health officials be allowed to go this far? The last time they attempted to interfere with reproductive freedom to a significant degree was during the twentieth-century eugenic movement, where they sterilized people without their consent to prevent them from passing on "defective" genes to their children. Regulating parents' genetic engineering decisions might be more acceptable, however, since it clearly would be less intrusive insofar as it did not entail surgically altering either the parents or their children. In fact, the measures needed to preserve the lineage bear a resemblance to what some experts have suggested might be needed in order to curb runaway population growth. In 2009, then *Fox News* broadcaster Glenn Beck accused John Holdren, President Obama's director of the White House Office of Science and Technology Policy, of advocating that sterilants be added to drinking water for this purpose.[27] This was typical Glenn Beck; what Holdren actually had said in a 1977 book he wrote with Paul and Anne Ehrlich was that there was no sign of a safe sterilizing agent on the horizon, and that "the risk of serious, unforeseen side effects, would, in our opinion, militate against the use of *any* such agent, even though this plan has the advantage of avoiding the need for socioeconomic pressures that might tend to discriminate against particular groups or penalize children" (emphasis in original).[28] Still, Holdren and the Ehrlichs did indeed say that "people should long ago have begun exploring, developing, and discussing all possible means of population control. But they did not, and time has nearly run out."[29] They also wrote that "several coercive proposals deserve discussion, mainly because some countries may ultimately have to resort to them unless current trends in birth rates are rapidly reversed by other means,"[30] and that while "compulsory control of family size is an unpalatable idea . . . , the alternatives may be much more horrifying."[31] This suggests that there might be some sentiment, at least among anti-engineering activists, in favor of even a draconian regulatory program for evolutionary engineering if circumstances were deemed to warrant it.

If the necessary degree of government control over reproductive decision making garnered enough popular support to become the law, would it pass constitutional muster? On the one hand, the Supreme Court has accorded governmental authorities a broad degree of discretion when they act to protect life. "Where it has a rational basis to act, and it does not impose an undue burden," the court declared in *Gonzalez v. Carhart*, the case in which it upheld so-called partial-birth abortions, "the State may use its regulatory power to bar certain procedures and substitute others, all in furtherance of its legitimate interests in regulating the medical profession in order to promote respect for life, including life of the unborn."[32] John Holdren, the director of President Obama's White House Office of Science and Technology Policy, argued in the book mentioned earlier that "compulsory population-control laws, even including laws requiring compulsory abortion, could be sustained under the existing Constitution if the population crisis became sufficiently severe to endanger the society,"[33] and that "if society's survival depended on having more children, women could be required to bear children, just as men can constitutionally be required to serve in the armed forces."[34] Holdren does not cite any authority for his expansive claims about the power of the state to control reproductive behavior, but if he is correct that laws to protect society can pass constitutional scrutiny, then the same ought to be true for laws that combat threats to the survival of the lineage.

On the other hand, when the Supreme Court refers to "life" in the *Gonzales* case, it clearly did not have in mind the human lineage. The intrusions on reproductive freedom that have been sanctioned by the court, such as bans on later-term and "partial-birth" abortions, have all been aimed at ostensibly protecting embryos, fetuses, children, and mothers, rather than at preserving something as abstract and remote as the future of humanity. Nor is it easy to predict how justices would vote on a lineage protection program based on their religious or ideological viewpoints. Ordinarily the conservatives are most willing to restrict reproductive choices to protect future persons. Yet more libertarian-minded justices would be disinclined to uphold an intrusive regulatory regime that limited people's ability to have children, even one that only applied to parents who wished to engineer their offspring. Liberal justices traditionally have favored the maximum amount of reproductive freedom for parents, but they also would be likely to be more sympathetic to evolutionary engineering and therefore perhaps more willing to regulate it so that it could proceed more safely and efficaciously. A highly regulated approach for reproductive choices that was adopted to protect downstream generations might become somewhat more acceptable to the public and the courts after people had gotten used to a first wave of programs that permitted the government to intervene to protect engineered children from harm, but even so, the fate of these government efforts in the courts is difficult to predict, at least so long as threats to the lineage seem remote.

So perhaps adequate protections for the lineage cannot be implemented in countries with a strong tradition of individual liberty like the United States, and many of their citizens may not survive or continue to reproduce in the event of a sudden, severe environmental challenge. But this does not necessarily mean that the human lineage is doomed. People in the Western democracies may become extinct, but the human lineage might persist so long as enough people managed to overcome extinction elsewhere. Whereas a future regulatory regime in which the government monitored and controlled reproductive decision making may be abhorrent to liberal societies, it may be acceptable in countries that have different political philosophies.

A number of countries, for instance, have laws against sex selection, including England, Australia, Canada, Germany, France, and South Korea, and it is prohibited by the Council of Europe's 1997 Convention on Human Rights and Biomedicine except to avoid a sex-related genetic disorder, that is, one that stems from mutations on either the X or Y sex chromosome.[35] In England, only private clinics that are not part of the National Health Service are allowed to perform sperm sorting for purposes of selecting the gender of IVF embryos.

India also has been no stranger to severe population control practices. After the government declared a state of emergency in the 1970s, it implemented a program of forced sterilization in the poorest parts of the country. One Indian state, Maharashtra, required men and women to be sterilized within six months of having their third child, and those who refused could be imprisoned for up to two years.[36] To prevent lawsuits against the health care workers who performed the operations, the law declared that the surgeries did not constitute serious harm, and the law also stated that they did not violate religious principles. People who practiced birth control, primarily the middle and upper classes, were permitted to opt for compulsory abortion rather than sterilization after their third child. Informants who ratted on their neighbors were entitled to rewards, as were the medical workers who completed the most procedures.[37] In 1976, government employees in New Delhi who had more than two children also were required to be sterilized and were denied public housing and other benefits.[38]

One technique used by the Indian government was to hold "family planning festivals." The Ehrlichs and Holdren provide the following description of one such event: "In one district during July 1971, over 60,000 vasectomies were performed at one festival. IUDs, condoms, and female sterilizations were also available. Greater than usual incentive payments and gifts were offered both to recipients and recruiters. The festival also included entertainment and cultural events. There was a great deal of publicity, and entertainers toured the surrounding countryside beforehand to attract people to the festival."[39]

This description makes these mass sterilization fairs sound like trips to the amusement park. In his novel *A Fine Balance*, however, Rohinton Mistry provides a very different portrayal of the Indian population control program and, in particular, how it caught up with Ishvar and his 18-year-old nephew Om, two members of an untouchable caste who return to their village to find Om a wife. Ishvar has just purchased a cotton candy for Om from a street vendor when a fleet of trucks pulls up and they are grabbed along with everyone else in the village market, hustled onto the trucks, and driven to a sterilization camp. Like a family planning festival, the camp is festooned with balloons and banners. But this is no party. When four people dragged from one truck start screaming, a doctor warns them to stop resisting because "if the knife slips it will harm you only." An elderly woman asks why they are wasting time on her, since she is long past child-bearing age. An official remarks that "targets have to be achieved," so the doctor is told to proceed. "These people often lie about their age," the official adds. "And appearances are deceptive. With their lifestyle, thirty can look like sixty, all shriveled by the sun." A woman on the truck hands her baby to Ishvar as she climbs down. Ishvar holds the baby, and as he thinks of the children Om will never have and tears roll down his cheeks, Om turns away. Ishvar "did not need to ask the reason," Mistry tells us. After Ishvar and Om are sterilized, a man in the recovery tent reveals that he previously had been sterilized of his own free will. "These animals did it on me today for the second time." "That's like executing a dead man," says Ishvar. "Don't they listen to anything?" "What to do, bhai," replies the man, "when educated people are behaving like savages. How do you talk to them? When the ones in power lose their reason there is no hope." While the operations are underway, an official tells the doctors there is no time to wait for their instruments to be completely sterilized. "He threatened to report them to higher authorities for lack of cooperation, promotions would be denied, salaries frozen."[40] As a result of being operated on with dirty instruments, Ishvar's legs become infected and have to be amputated. Whereas before they were apprentice tailors, he and Om now have no choice but to become street beggars.

The country that currently goes the farthest in terms of controlling its citizens' reproductive behavior is not India, however, but China. Chapter 6 described elements of the Chinese program that are designed to prevent the birth of children with disabilities, but the better-known aspect of China's regulation of its citizen's reproductive decisions is the government's one-child policy. Before the policy was adopted, China had gone through a succession of flip-flops over population control. When the Communists originally gained power, they regarded a larger population as an important measure of national strength, so they banned imports of contraceptives. But birth control gained official sanction after the country was battered by food shortages in the early 1950s; Americans who are old enough may remember their parents encouraging them to eat everything put on their plates by reminding them of "all the starving children in China." Then came the Great Leap forward, and all of a sudden population growth was desirable again. In 1958, for instance, the secretary of the Communist Youth League remarked that "the force of 600 million liberated people is tens of thousands of times stronger than a nuclear explosion."[41] Later, population control was imposed once more after famine consumed an estimated 30 million lives. "In some of the cities," the Ehrlichs and Holdren wrote in 1977, "low birth rates have been so enthusiastically adopted as a goal that neighborhoods collectively decide how many births will be allowed each year and award the privilege to 'deserving couples.'"[42] The results were lackluster, however, which led to the initiation in 1979 of the one-child policy, primarily for the majority Han ethnic group. Couples who have more than one child face severe fines, and there have been reports of involuntary abortions and sterilizations.[43] The Chinese government contends that the policy has prevented 400 million births.[44]

The prospect that states such as China that are willing to impose significant controls on reproductive freedom would be the ones to salvage the human lineage is, at the very least, ironic. In his 1992 book *The End of History and the Last Man*, Frances Fukuyama maintains that "the universalization of Western liberal democracy" is "the final form of human government."[45] But if liberal democracies cannot find a suitable balance between liberty and long-term survival, then as the saying goes, humans may end up all speaking Chinese.

Epilogue

Jews from Eastern Europe have difficulty tracing their families back very far. Unlike Gentiles, we have no family bibles in which to engrave the names of our forbearers, and whatever records the synagogues kept were burned by the Nazis. I spent part of my childhood summers at my grandmother's on my mother's side, but my father's parents died when I was a baby, and I never learned the identity or fate of my maternal grandfather. So sometimes I play a little mind game and invent a notional set of ancestors. I go back in time from one to the other, each one getting a third of a century and a brief moment of my imagination. I spend a few seconds giving them faces, clothing, and a bit of a narrative, and then I jump to the preceding one.

If you have never done this yourself, you should give it a try. See how far back you can get.

I am not able to keep two lineages in mind at the same time, so I only follow my father's family. Even so, it is an awesome journey. Only sixty people stand between me and the ancestor who lived at the time of Christ. Go back another 780 and there's the first one to grow crops. Thirty thousand or so farther is the first one to make fire, another 270,000, the last one to swing through trees. I am the living embodiment of every one of those beings, and I feel a responsibility to them all.

Each of us stands not only at the end of such a long line of life but at the beginning of another, one that vanishes into a distant, potentially infinite future. We owe a responsibility to the members of that line too. We are *their* ancestors.

Our descendants will encounter many great challenges. They will have to skirt global catastrophe, perhaps frequently. By the time the sun dies, they will have to have colonized space, where they no doubt will meet their first intelligent aliens. Yet humanity faces a great test now as well. Rather than just passing genes on to our offspring the way those before us did, we are acquiring the technological wherewithal to reconstruct those genes. If we botch it, children will suffer, the lineage may die out, and that will be that. If we succeed, we will earn the gratitude of our descendants. It seems to me that we owe it to all those ancestors and to all those potential descendants to get it right. We also owe it to each other. After all, many evolutionary biologists happen to agree with creationists that, at some point in the past, all of our ancestors were the same person.

NOTES

Introduction

1. Gallop Poll, "Evolution, Creationism, Intelligent Design," www.gallup .com/poll/21814/evolution-creationism-intelligent-design.aspx.

2. Richard Dawkins, "God's Utility Function," *Scientific American*, Nov. 1995, 85.

3. "Marriageable Size," *Healthy Weight Journal* 11, no. 3 (1997): 46.

4. C. Beckwith, "Niger's Wodaabe, 'People of the Taboo,'" *National Geographic* 164, no. 4 (1983): 82.

5. Mollie H. Chen, "Welcome to the Ball: A Look inside Debutante Culture," *Harvard Crimson*, May 2, 2002, www.thecrimson.com/article/2002/5/2/welcome -to-the-ball-kathryn-a/.

6. www.eharmony.com/why/dimensions.

7. Natasha Singer, "Better Loving through Chemistry," *New York Times*, Feb. 7, 2010, www.nytimes.com/2010/02/07/business/07stream.html.

8. Ibid.

9. 23andMe, "23andMe Democratizes Personal Genetics," Sept. 9, 2008, www.23andMe/about/press/20080909b/.

10. Fertility Alternatives, www.fertilityalternatives.com/eggdonors.html#16.

11. Jason Palmer, "Glowing Monkeys 'to Aid Research,'" *BBC News*, May 27, 2009, http://news.bbc.co.uk/2/hi/8070252.stm.

12. Lewis Thomas, *The Lives of a Cell: Notes of a Biology Watcher* (New York: Penguin, 1978), 37.

13. Peter Ward, *Future Evolution: An Illuminated History of Life to Come* (New York: Times Books, 2001), 146.

14. Julian Huxley, "Transhumanism," in *New Bottles for New Wine* (London: Chatto and Windus, 1957), 13–17.

15. Joseph Fletcher, *The Ethics of Genetic Control* (Buffalo, NY: Prometheus Books, 1988), 158.

16. Gregory Stock, "Germinal Choice Technology and the Human Future," *Reproductive Biomedicine Online* 10 (2005): 27, 34.

17. Fletcher, *Ethics of Genetic Control*, 158.

18. Richard Dawkins, *The Selfish Gene* (Oxford: Oxford University Press, 1976), 215.

19. Julian Savulescu, "Human-Animal Transgenesis and Chimeras Might Be

an Expression of Our Humanity," *American Journal of Bioethics* 3, no. 3 (2003): 22–25.

20. Nick Bostrom, "In Defence of Posthuman Dignity," *Bioethics* 19, no. 3 (2005): 202–14.

21. Stock, "Germinal Choice Technology," 34.

22. "Address of His Holiness Benedict XVI to the Participants in the General Assembly of the Pontifical Academy for Life," Feb. 24, 2007, www.vatican.va/holy_father/benedict_xvi/speeches/2007/february/documents/hf_ben-xvi_spe_20070224_academy-life_en.html.

23. Leon Kass, "Ageless Bodies, Happy Souls: Biotechnology and the Pursuit of Perfection," *New Atlantis* 1 (2003): 9–28.

24. JASON 2010, *The $100 Genome: Implications for DoD* (Report No. JSR-10-100, Dec. 2010), www.fas.org/irp/agency/dod/jason/hundred.pdf.

25. Francis Collins, "The Promise and Peril of the Genomic Revolution" (Inamori Prize Lecture, Cleveland, OH, Sept. 4, 2008).

26. Statement of Theodore Friedman, First Gene Therapy Policy Conference, National Institutes of Health, Sept. 11, 1997.

CHAPTER ONE. *Visions of Heaven and Hell*

1. Simon Young, www.simonyoungmassage.co.uk/.

2. Simon Young, www.designerevolution.net/.

3. Ibid.

4. Simon Young, *Designer Evolution: A Transhumanist Manifesto* (Amherst, NY: Prometheus Books, 2006), 37.

5. Young, www.designerevolution.net/.

6. Gregory Stock, *Redesigning Humans: Our Inevitable Genetic Future* (Boston: Houghton Mifflin, 2002), 39.

7. James Hughes, *Citizen Cyborg: Why Democratic Societies Must Respond to the Redesigned Human of the Future* (Cambridge, MA: Westview Press, 2004), 11–12.

8. Lee M. Silver, *Remaking Eden: Cloning and Beyond in a Brave New World* (New York: Avon Books, 1997), 237–38.

9. Ibid.

10. ·Stock, *Redesigning Humans*, 79.

11. John Harris, *Enhancing Evolution: The Ethical Case for Making Better People* (Princeton, NJ: Princeton University Press, 2007), 12.

12. Ibid.

13. Joel Garreau, *Radical Evolution: The Promise and Peril of Enhancing Our Minds, Our Bodies—and What It Means to Be Human* (New York: Doubleday, 2005), 7–8.

14 Silver, *Remaking Eden*, 237–38.

15. Ramez Naam, *More Than Human: Embracing the Promise of Biological Enhancement* (New York: Broadway Books, 2005), 55.

16. Hughes, *Citizen Cyborg*, 47.

17. Young, *Designer Evolution*, 37.

18. Ibid.

19. American Society of Plastic Surgeons, 2010 Report of the 2009 Statistics, National Clearinghouse of Plastic Surgery Statistics, 6, www.plasticsurgery.org/Documents/news-resources/statistics/2009-statistics/2009-cosmetic-reconstructive-plastic-surgery-minimally-invasive-statistics.pdf.

20. Ibid., 12.

21. Michael A. Hiltzik, "Presumed Guilty: Athletes' Unbeatable Foe," *Los Angeles Times*, Dec. 10, 2006, A1.

22. World Anti-Doping Agency, www.wada-ama.org/en/dynamic.ch2?pageCategory.id=264.

23. International Paralympic Committee Press Release 2004, www.paralympic.org/release/Main_Sections_Menu/News/Press_Releases/2004_09_26_a.html.

24. World Anti-Doping Agency, "Questions and Answers on EPO Detection," www.wada-ama.org/en/Resources/Q-and-A/Q-A-EPO-Detection/.

25. L. A. Whittemore et al., "Inhibition of Myostatin in Adult Mice Increases Skeletal Muscle Mass and Strength," *Biochemical and Biophysical Research Communications* 300, no. 4 (2003): 965–71.

26. David Sadava et al., *Life: The Science of Biology*, 7th ed. (New York: Macmillan, 2009), 168.

27. Garreau, *Radical Evolution*.

28. Silver, *Remaking Eden*, 137–38.

29. Garreau, *Radical Evolution*, 7–8.

30. Julian Savulescu, "The Human Prejudice and the Moral Status of Enhanced Beings: What Do We Owe the Gods?," in *Human Enhancement*, ed. Julian Savulescu and Nick Bostrom (Oxford: Oxford University Press, 2009), 213–14.

31. Silver, *Remaking Eden*, 237–38.

32. Ibid.

33. Young, *Designer Evolution*, 37.

34. Garreau, *Radical Evolution*, 7–8.

35. Hughes, *Citizen Cyborg*, 39.

36. "Tracking Down Genes for Intelligence," Netherlands Organisation for Scientific Research (NWO), Mar. 15, 2010, www.nwo.nl/nwohome.nsf/pages/NWOP_83KCW5_Eng.

37. Y. Tang et al., "Genetic Enhancement of Learning and Memory in Mice," *Nature* 401 (1999): 63–69.

38. Katherine E. Burdick et al., "Genetic Variation in DTNBP1 Influences General Cognitive Ability," *Human Molecular Genetics* 15, no. 10 (2006): 1563–68; M. F. Gosso et al., "Common Variants Underlying Cognitive Ability: Further Evidence for Association between the SNAP-25 Gene and Cognition Using a Family-Based Study in Two Independent Dutch Cohorts," *Genes, Brain and Behavior* 7 (2008): 355–64.

39. Mauro Costa-Mattioli et al., "Phosphorylation Bidirectionally Regulates

the Switch from Short- to Long-Term Synaptic Plasticity and Memory," *Cell* 129, no. 1 (2007): 195–206.

40. José A. Apud et al., "Tolcapone Improves Cognition and Cortical Information Processing in Normal Human Subjects," *Neuropsychopharmacology* 32 (2007): 1011–20.

41. JASON 2010, *The $100 Genome: Implications for DoD* (Report No. JSR-10-100, Dec. 2010), www.fas.org/irp/agency/dod/jason/hundred.pdf.

42. Stock, *Redesigning Humans*, 213.

43. Ibid., 63.

44. Garreau, *Radical Evolution*, 7–8.

45. Robin Hanson, "Enhancing Our Truth Orientation," in Savulescu and Bostrom, *Human Enhancement*, 366.

46. Stock, *Redesigning Humans*, 105; Silver, *Remaking Eden*, 237–38.

47. Stock, *Redesigning Humans*, 105.

48. Savulescu, "Human Prejudice," 213, 287.

49. Naam, *More Than Human*, 237–38.

50. Ray Kurzweil, *The Singularity Is Near: When Humans Transcend Biology* (New York: Viking, 2005), 11–12; Naam, *More Than Human*, 30.

51. Kurzweil, *Singularity Is Near*, 11–12.

52. Hughes, *Citizen Cyborg*, 19–20.

53. Ibid.

54. Ibid., 48.

55. Naam, *More Than Human*, 55.

56. Garreau, *Radical Evolution*, 242.

57. Ibid.

58. Harris, *Enhancing Evolution*, 59.

59. "About the A4M," www.worldhealth.net/about-a4m/.

60. Duff Wilson, "Aging: Disease or Business Opportunity?," *New York Times*, Apr. 15, 2007, sec. 3, 1.

61. 21 U.S.C. § 333(h).

62. Leonard Hayflick, "Anti-Aging Medicine: Hype, Hope, and Reality," *Generations* 25 (2001–2): 20–26.

63. M. R. Blackman et al., "Growth Hormone and Sex Steroid Administration in Healthy Aged Women and Men: A Randomized Controlled Trial," *Journal of the American Medical Association* 288 (2001–2): 282–92; H. Janssens and D. M. O. I. Vanderschueren, "Endocrinological Aspects of Aging in Men: Is Hormone Replacement of Benefit?," *European Journal of Obstetrics and Gynecology and Reproductive Biology* 92 (2000): 7–12.

64. J. M. Chan et al., "Plasma Insulin-Like Growth Factor I and Prostate Cancer Risk: A Prospective Study," *Science* 279 (1998): 563–66.

65. S. A. Shumaker et al., "Estrogen plus Progestin and the Incidence of Dementia and Mild Cognitive Impairment in Postmenopausal Women—the Women's Health Initiative Memory Study: A Randomized Controlled Trial," *Journal of the American Medical Association* 289 (2003): 2651–62.

66. Writing Group for the Women's Health Initiative Investigators, "Risks and Benefits of Estrogen plus Progestin in Healthy Postmenopausal Women: Principal Results from the Women's Health Initiative Randomized Controlled Trial," *Journal of the American Medical Association* 288 (2002): 321–33.

67. D. Sinclair and A. L. Komaroff, "Can We Slow Aging?," *Newsweek* 148 (2006): 80–84.

68. Harris, *Enhancing Evolution*, 59.

69. Young, *Designer Evolution*, 42.

70. Ibid., 49.

71. Anita Silvers, "Future Perfect: The Transhumanist Quest to Defeat Death," *New Humanist* 121, no. 6 (2006), http://newhumanist.org.uk/1497/future -perfect-the-transhumanist-quest-to-defeat-death.

72. Mandy Kendrick, "Anti-Aging Pill Targets Telomeres at the Ends of Chromosomes," *Scientific American*, Aug. 17, 2009, www.scientificamerican.com/ article.cfm?id=anti-aging-pill-targets-telomeres&print=true.

73. http://telonauts.com/wordpress/.

74. Hughes, *Citizen Cyborg*, 27.

75. Ibid., 38–39.

76. Ibid.

77. Savulescu, "Human Prejudice," 212.

78. "Secrets of the Centenarians," *Howard Hughes Medical Institute Bulletin* 17, no. 1 (2004), www.hhmi.org/bulletin/spring2004/centenarians/centenarians 2.html.

79. Almut Nebel et al., "No Association between Microsomal Triglyceride Transfer Protein (*MTP*) Haplotype and Longevity in Humans," *Proceedings of the National Academy of Sciences* 102, no. 22 (2005): 7906–9.

80. Friederike Flachsbart et al., "Association of *FOXO3A* Variation with Human Longevity Confirmed in German Centenarians," *Proceedings of the National Academy of Sciences* 106, no. 8 (2009): 2700–2705.

81. James Burnett, "Take This Pill and Live Forever," *Boston Magazine*, Apr. 2003, www.bostonmagazine.com/articles/take_this_pill_and_live_forever/.

82. Kurzweil, *Singularity Is Near*, 212.

83. "Mormon Tranhmanist Affirmation," http://transfigurism.org/pages/about/ mormon-transhumanist-affirmation/.

84. "Humanity+ Transhumanist FAQ," http://humanityplus.org/learn/trans humanist-faq/#answer_46.

85. Simon Young, May 10, 2006, www.warrenellis.com/?p=2464.

86. Robert Heinlein, *The Eyes of Heisenberg* (New York: Tor, 1966).

87. Lois McMaster Bujold, *Free Falling* (Riverdale, NY: Baen, 1988), 6.

88. Tim Dirks, "AMC Filmsite," www.filmsite.org/blad4.html.

89. Michael Crichton, *Next* (New York: HarperCollins, 2006), 4–5.

90. Ibid., 3.

91. Robert Heinlein, *Friday* (New York: Ballantine Books, 1982).

92. Nancy Kress, *Beggars in Spain* (New York: HarperCollins, 1993), 12.

93. Ibid., 51.

94. Ibid.

95. Ibid., 74.

96. Presumably the Sleepless are homozygous for the Sleepless trait, that is, they each have two copies of the Sleepless genes. Otherwise, the children they conceived with Sleepers would only have a 50–50 chance of also being Sleepless.

97. Ibid., 75.

98. Ibid., 120.

99. Kress, *Beggars in Spain.*

100. Ibid., 245.

101. Aldous Huxley, *Brave New World*, www.huxley.net/soma/somaquote .html.

102. Margaret Atwood, *Oryx and Crake* (New York: Anchor, 2003), 55.

103. Ibid., 56–57.

104. Ibid., 211.

105. Ibid., 202.

106. Ibid., 203.

107. Ibid., 303.

108. Ibid., 8.

109. Ibid., 305.

110. Ibid., 164.

111. Ibid., 165.

112. Ibid., 303.

113. Ibid., 306.

114. Kress, *Beggars in Spain*, 7.

115. Atwood, *Oryx and Crake*, 55.

116. Ibid., 156.

117. Ibid., 295.

118. Crichton, *Next*, 169.

CHAPTER TWO. *Thinking about the Unthinkable*

1. Cass R. Sunstein, *Laws of Fear: Beyond the Precautionary Principle* (New York: Cambridge University Press, 2005).

2. Cass R. Sunstein, *Worst-Case Scenarios* (Cambridge, MA: Harvard University Press, 2007).

3. Richard A. Posner, *Catastrophe: Risk and Response* (New York: Oxford University Press, 2004).

4. Ibid., 176.

5. Ibid., 179–80.

6. National Institutes of Health, "ELSI Planning and Evaluation History," www.genome.gov/10001754.

7. Congressional Research Service, "The FDA FY2009 Budget Request, CRS-4," http://digital.library.unt.edu/govdocs/crs/permalink/meta-crs-10745:1.

8. David Hawkins, *Manhattan District History, Project Y, The Los Alamos Project*, vol. 1 (Los Alamos Scientific Laboratory), quoted in Richard Rhodes, *The Making of the Atomic Bomb* (New York: Simon and Schuster, 1986), 419.

9. U.S. Department of Energy, The Manhattan Project, www.cfo.doe.gov/me70/manhattan/trinity.htm.

10. Rhodes, *Making of the Atomic Bomb*, 405.

11. Joseph I. Kapusta, "Accelerator Disaster Scenarios, the Unabomber, and Scientific Risks," *Physics in Perspective* 10 (2008): 163, 166.

12. Ibid., 169.

13. Ibid., 173.

14. Ibid., 175.

15. European Organization for Nuclear Research, "The Safety of the LHC," http://public.web.cern.ch/public/en/LHc/Safety-en.html.

16. Martin Rees, *Our Final Hour: A Scientist's Warning: How Terror, Error, and Environmental Disaster Threaten Humankind's Future in this Century—on Earth and Beyond* (New York: Basic Books, 2003), 120–21, quoted in Posner, *Catastrophe*, 30.

17. Robert L. Jaffe, Wit Busza, and Frank Wilczek, "Review of Speculative 'Disaster Scenarios' at RHIC," *Reviews of Modern Physics* 72 (2000): 1125.

18. See Kapusta, "Accelerator Disaster Scenarios," 176.

19. CERN stands for Conseil Européen pour la Recherche Nucléaire.

20. Frank Close, "Ions in the Fire: Particle Physicists Are on Collision Course to Recreate Conditions at the Birth of the Universe," *Guardian*, Apr. 29, 1999, 10.

21. See, e.g., "France Builds Doomsday Machine," www.misunderstooduniverse.com/France_Builds_Doomsday_Machine.htm; "Stop the LHC—Until We Know It's Safe," www.lhcdefense.org/lhc_experts9.php.

22. "CERN: History," http://lhc.web.cern.ch/lhc/general/history.htm.

23. Kevin Mayhood, "What Do You Get When You Give $8 Billion to 8,000 Physicists? . . . A Science Project That May Rock the Universe," *Columbus Dispatch*, Sept. 23, 2008, 04B.

24. "Angels and Demons (CERN)," http://public.web.cern.ch/Public/en/Spotlight/SpotlightAandD-en.html.

25. Sancho v. Dept. of Energy, 378 F. Supp. 2d 1258 (D. Haw. 2008).

26. Dennis Overbye, "Suit to Halt Big Collider in Europe Is Dismissed," *New York Times*, Sept. 30, 2008, 21.

27. Paul Wagonseil, "Lawsuit: Huge Atom Smasher Could Destroy World," www.foxnews.com/story/0,2933,342854,00/html.

28. *Sancho*, 1267, 1269.

29. Ibid., 1269.

30. See J.-P. Blaizot et al., "Study of Potentially Dangerous Events during Heavy-Ion Collisions at the LHC: Report of the LHC Safety Group" (CERN 2003), http://cdsweb.cern.ch/record/613175?ln=pl; Benjamin Koch, Marcus Bleicher, and Horst Stoecker, "Exclusion of Black Hole Disaster Scenarios at the LHC," *Physics Letters* 71 (2009): 672; Marco Cavaglia, Saurya Das, and Roy

Maartens, "Will We Observe Black Holes at LHC?," *Classical and Quantum Gravity* 20 (2003): L205.

31. Dennis Overbye, "Ah Spring! Swallows, Baseball, Colliding Protons," *New York Times*, Sept. 24, 2008, 7.

32. Adam Gabbatt, "Big Bang Goes Phut as Bird Drops Baguette into Cern Machinery," *Guardian*, Nov. 7, 2009, 25.

33. David A. Jackson, Robert H. Symons, and Paul Berg, "Biochemical Method for Inserting New Genetic Information into DNA of Simian Virus 40: Circular SV40 DNA Molecules Containing Lambda Phage Genes and the Galactose Operon of *Escherichia coli*," *Proceedings of the National Academy of Sciences* 69 (1972): 2904.

34. Diamond v. Chakrabarty, 447 U.S. 303 (1980).

35. Jon Van, "In Oil-Spill Cleanups, Major Tool Off-Limits," *Chicago Tribune*, June 18, 1989, C3.

36. Malcolm Brown, "Researchers Enlist Bacteria to Do Battle with Oil Spill," *New York Times*, May 23, 1989, C1.

37. N. Ferrer-Miralles et al., "Microbial Factories for Recombinant Pharmaceuticals," *Microbial Cell Factories* 8 (2009): 17.

38. National Cancer Institute, "Simian Virus 40 and Human Cancer: Questions and Answers," www.cancer.gov/newscenter/sv40.

39. Paul Berg et al., "Potential Biohazards of Recombinant DNA Molecules," *Science* 185 (1974): 303.

40. Alan Lightman, *The Discoveries* (New York: Pantheon Books, 2005).

41. Posner, *Catastrophe*, 208, 202.

42. Ibid., 208.

43. Ibid., 203.

44. Blaizot et al., "Study of Potentially Dangerous Events," 12.

45. Enita A. Williams, "Good, Better, Best: The Human Quest for Enhancement: Summary Report of an Invitational Workshop Convened by the Scientific Freedom, Responsibility and Law Program of the American Association for the Advancement of Science" (AAAS, 2006).

46. Project on Emerging Nanotechnologies, www.nanotechproject.org/inven tories/consumer/.

47. Rick Weiss, "Groups Petition EPA to Ban Nanosilver in Consumer Goods," *Washington Post*, May 2, 2008, A04.

48. Royal Commission on Environmental Pollution, "Novel Materials in the Environment: The Case of Nanotechnology," www.rcep.org.uk/reports/27-novel %20materials/documents/NovelMaterialsreport.pdf.

49. Atsuya Takagi et al., "Induction of Mesothelioma in p53+/– Mouse by Intraperitoneal Application of Multi-wall Carbon Nanotube," *Journal of Toxicological Sciences* 33, no. 1 (2008): 105.

50. Friends of the Earth, "Nano and Biocidal Silver," June 2009, www.foe .org/sites/default/files/Nano-silverReport_US.pdf.

51. U.S. Environmental Protection Agency, "Significant New Use Rules on Certain Chemical Substances," *Federal Register* 74 (2009): 29982–98.

52. K. Eric Drexler, *Engines of Creation: The Coming Era of Nanotechnology* (New York: Anchor Press / Doubleday, 1986).

53. Mitchell Martin, "Technology's Little-Heeded Prophet," *International Herald Tribune*, Oct. 23, 2000, 12.

54. Michael Crichton, *Prey* (New York: HarperCollins, 2002).

55. Michael Brunton, "Little Worries," www.time.com/time/magazine/article/0,9171,449458,00.html.

56. Charles, Prince of Wales, "Small Is Hazardous," *Independent on Sunday*, July 11, 2004, 25.

57. Richard E. Smalley, "Of Chemistry, Love, and Nanobots," *Scientific American* 285, no. 3 (2001): 76.

58. Dan Charles, "Proposed Regulatory Czar Has Long and Perplexing Track Record," *Science* 23, no. 5913 (2009): 452–53.

59. International Risk Governance Council, "Nanotech Risk Governance: Recommendations for a Global, Coordinated Approach to the Governance of Potential Risks," www.irgc.org/IMG/pdf/PB_nanoFINAL2_2_.pdf.

60. U.S. Department of Transportation, National Highway Traffic Safety Administration, Federal Motor Vehicle Safety Standards; Air Brake Systems (Docket No. NHTSA-2009-0083), www.nhtsa.gov/staticfiles/DOT/NHTSA/Rulemaking/Rules/Associated%20Files/121_Stopping_Distance_FR.pdf.

CHAPTER THREE. *Physical Harm to Children*

1. David Suzuki and Peter Knudtson, *Genethics: The Clash between the New Genetics and Human Values* (Cambridge, MA: Harvard University Press, 1989), 33.

2. Noel Kingsbury, *Hybrid: The History and Science of Plant Breeding* (Chicago: University of Chicago Press, 2009), 74.

3. Peter Ward, *Future Evolution: An Illuminated History of Life to Come* (New York: Times Books, 2001), 108.

4. Ibid., 107.

5. J. P. Gibson, W. Ayalew, and O. Hanotte, "Measures of Diversity as Inputs for Decisions in Conservation of Livestock Genetic Resources," in *Managing Biodiversity in Agricultural Ecosystems*, ed. D. I. Jarvis, C. Padoch, and H. D. Cooper (New York: Columbia University Press, 2006), 117–40.

6. I. Hoffman, "Management of Farm Animal Genetic Resources: Change and Interaction," in Jarvis, Padoch, and Cooper, *Managing Biodiversity*, 146.

7. Ibid.

8. Robert T. Francoeur, "We Can—We Must: Reflections on the Technological Imperative," *Journal of Theological Studies* 33 (1972): 428–39.

9. U.S. FDA Center for Food Safety and Applied Nutrition, "Backgrounder: Biotechnology of Food," May 18, 1995, www.foodsafety.gov/~lrd/biotechn.html.

10. Gregory N. Mandel, "Gaps, Inexperience, Inconsistencies, and Overlaps: Crisis in the Regulation of Genetically Modified Plants and Animals," *William and Mary Law Review* 45 (2004): 2177.

11. Angela Hall, "Suzuki Warns against Hastily Accepting GMOs," *Leader-Post* (Canada), Apr. 26, 2005.

12. Hugh S. Lehman, "Dangers from Consumption of Foods Containing Transplanted DNA," www.sierraclub.org/biotech/whatsnew/whatsnew_2006-04 -12.asp.

13. David R. Schubert, "The Problem with Nutritionally Enhanced Plants," *Journal of Medicinal Food* 11, no. 4 (2008): 601–5.

14. Mandel, "Gaps, Inexperience, Inconsistencies, and Overlaps," 2205–6.

15. U.S. Department of Energy, Human Genome Project Information, "Genetically Modified Foods and Organisms," www.ornl.gov/sci/techresources/Human _Genome/elsi/gmfood.shtml.

16. Mandel, "Gaps, Inexperience, Inconsistencies, and Overlaps," 2177.

17. Ibid., 2181.

18. Ibid., 2184.

19. Ibid., 2183n49.

20. Ibid., 2187.

21. U.S. Department of Energy, "Genetically Modified Foods."

22. Stephen Sawicki, "Super Trees," *New York Times*, Nov. 21, 2004, 14CN.

23. Ibid.

24. Sam Aola Ooko, "Genetically-Engineered Tobacco Bio-sensor to Detect Landmines," *EcoWorldly*, July 29, 2008, http://ecoworldly.com/2008/07/29/ genetically-engineered-tobacco-bio-sensor-to-detect-landmines/.

25. Mandel, "Gaps, Inexperience, Inconsistencies, and Overlaps," 2189.

26. Ward, *Future Evolution*, 110.

27. Doug Farquhar and Liz Meyer, "State Authority to Regulate Biotechnology under the Federal Coordinated Framework," *Drake Journal of Agricultural Law* 12 (2007): 448–49.

28. Ward, *Future Evolution*, 111.

29. Francoeur, "We Can—We Must," 429.

30. European Society for Reproduction and Embryology, "Three Million Babies Have Been Born Using Assisted Reproductive Technologies," *Women's Health Law Weekly*, July 16, 2006, cited in Judith Darr, "Embryonic Genetics," *St. Louis University Journal of Health Law & Policy* 2, no. 1 (2008): 81.

31. A. H. Handyside et al., "Biopsy of Human Preimplantation Embryos and Sexing by DNA Amplification," *Lancet* 1, no. 8634 (1989): 347–49.

32. Gregory Stock, *Redesigning Humans: Our Inevitable Genetic Future* (New York: Houghton Mifflin, 2002).

33. Ann Hallum, "Disability and the Transition to Adulthood: Issues for the Disabled Child, the Family, and the Pediatrician," *Current Problems in Pediatrics* 25 (1995): 12–50.

34. Ibid.

35. www.cysticfibrosismedicine.com/htmldocs/CFText/historyof.htm.

36. Murray G. Baron and Wendy M. Book, "Congenital Heart Disease in the Adult," *Radiologic Clinics of North America* 42 (2004): 675–90.

37. Stock, *Redesigning Humans*, 38–39.

38. L. A. Whittemore et al., "Inhibition of Myostatin in Adult Mice Increases Skeletal Muscle Mass and Strength," *Biochemical and Biophysical Research Communications* 300, no. 4 (2003): 965–71.

39. Ward, *Future Evolution*, 147.

40. J. A. Byrne et al., "Producing Primate Embryonic Stem Cells by Somatic Cell Nuclear Transfer," *Nature* 450 (2007): 497–502.

41. S. B. Christian and P. Sandøe, "Bioethics: Limits to the Interference with Life," *Animal Reproduction Science* 60–61 (2000): 15–29, 18–19.

42. National Research Council, *Animal Biotechnology: Research-Based Concerns* (Washington, DC: National Academies Press, 2002), 98.

43. Ibid.

44. Robert Taylor, "Superhumans: Like It or Not, in a Few Short Years We'll Have the Power to Control Our Own Evolution," *New Scientist*, Oct. 1, 1998, www.geneticsandsociety.org/article.php?id=129.

45. Christian and Sandøe, "Bioethics," 19.

46. Leroy Walters and J. G. Palmer, *The Ethics of Human Gene Therapy* (New York: Oxford University Press, 1997), cited in Harry Adams, "A Human Germline Modification Scale," *Journal of Law, Medicine & Ethics* 32 (2004): 164, 167.

47. Ward, *Future Evolution*, 105.

48. Ibid.

49. Ibid.

50. Rick Weiss, "Death Points to Risks in Research: One Woman's Experience in Gene Therapy Trial Highlights Weaknesses in the Patient Safety Net," *Washington Post*, Aug. 6, 2007, A01.

51. Michael Crichton, *Next* (New York: HarperCollins, 2006), 330–31.

52. Sheryl Gay Stolberg, "The Biotech Death of Jesse Gelsinger," *New York Times Sunday Magazine*, Nov. 28, 1999, 137.

53. Deborah Nelson and Rick Weiss, "NIH Not Told of Deaths in Gene Studies: Researchers, Companies Kept Agency in the Dark," *Washington Post*, Nov. 3, 1999, A01.

54. Stolberg, "Biotech Death," 137.

55. T. Gregory, "Clinical Applications of Molecular Medicine," *Patient Care* 32 (1998): 18, cited in Emily Marden and Dorothy Nelkin, "Displaced Agendas: Current Regulatory Strategies for Germline Gene Therapy," *McGill Law Journal* 45 (2000): 461, 470.

56. Marden and Nelkin, "Displaced Agendas," 468.

57. A. K. Ludwig et al., "Post-neonatal Health and Development of Children Born after Assisted Reproduction: A Systematic Review of Controlled Studies," *European Journal of Obstetrics & Gynecology and Reproductive Biology* 127 (2006): 3–25.

58. T. A. Manolio, "Genomic Medicine: Genomewide Association Studies and Assessment of the Risk of Disease," *New England Journal of Medicine* 363, no. 2 (2010): 166–76.

59. Ernst Mayr, *What Evolution Is* (New York: Basic Books, 2001).

60. W. French Anderson, "Human Gene Therapy: Why Draw a Line?," *Journal of Medicine and Philosophy* 14 (1989): 681–93.

61. Marc Lappé, "Moral Obligations and the Fallacies of 'Genetic Control,'" *Journal of Theological Studies* 33, no. 3 (1972): 411–27.

62. Ibid.

63. Nicholas Agar, *Liberal Eugenics: In Defense of Human Enhancement* (Malden, MA: Blackwell, 2004), 162.

64. Mark A. Rothstein, Yu Cai, and Gary E. Marchant, "The Ghost in Our Genes: Legal and Ethical Implications of Epigenetics," *Health Matrix* 19, no. 1 (2009): 7.

65. Susannah Baruch et al., "Human Germline Genetic Modifications: Issues and Options for Policymakers" (Genetics and Public Policy Center, 2005), 21.

66. Nick Bostrom and Anders Sandberg, "The Wisdom of Nature: An Evolutionary Heuristic for Human Enhancement," in *Human Enhancement*, ed. Julian Savulescu and Nick Bostrom (Oxford: Oxford University Press, 2009), 383.

67. Stuart Weiner, Janet Monge, and Alan Mann, "Bipedalism and Parturition: An Evolutionary Imperative for Cesarean Delivery?," *Clincs in Perinatology* 35, no. 3 (2008): 469–78.

68. M. M. Abitbol, *Birth and Human Evolution: Anatomical and Obstetrical Mechanics in Primates* (Westport, CT: Bergin and Garvey, 1996).

69. Ernst Mayr, *What Evolution Is* (New York: Basic Books, 2001), 248–49.

70. Charles E. Denk and Kathryn P. Aveni, "Surveillance of Maternal Peripartum Complications following Cesarean Section, New Jersey 1997–2005" (New Jersey Department of Health and Senior Services, 2009), www.nj.gov/health/fhs/professional/documents/maternal_complications_following_cesarean.pdf.

CHAPTER FOUR. *Psychosocial Harm to Children*

1. Emily Marden and Dorothy Nelkin, "Displaced Agendas: Current Regulatory Strategies for Germline Gene Therapy," *McGill Law Journal* 45 (2000): 461, 471.

2. Council for Responsible Genetics, "Position Paper on Human Germ Line Manipulation," *Human Therapy* 4 (1992): 35–37.

3. Ray Moseley, "Maintaining the Somatic/Germ Line Distinction: Some Ethical Drawbacks," *Journal of Medicine and Philosophy* 16 (1991): 641–47.

4. Marc Lappé, "Ethical Issues in Manipulating the Human Germline," *Journal of Medicine and Philosophy* 16 (1991): 621–39.

5. Ian G. Barbour, *Ethics in an Age of Technology: The Gifford Lectures 1989–91*, vol. 2 (1992), 197, cited by Michael J. Reiss, "What Sort of People Do We

Want? The Ethics of Changing People through Genetic Engineering," *Notre Dame Journal of Law, Ethics & Public Policy* 13 (1999): 63n66.

6. Nicholas Agar, *Liberal Eugenics: In Defense of Human Enhancement* (Malden, MA: Blackwell, 2004), 124.

7. Reiss, "What Sort of People?"

8. Marie-Noëlle Ganry-Tardy, "Watching Prodigies for the Dark Side," *Scientific American Mind*, Mar. 24, 2005, www.scientificamerican.com/article.cfm?id=watching-prodigies-for-th&print=true.

9. Laura Smith-Spark, "Growing Pains of a Child Prodigy," *BBC News*, May 27, 2006, http://news.bbc.co.uk/2/hi/americas/5009656.stm.

10. Mark S. Frankel and Audrey R. Chapman, *Human Inheritable Genetic Modifications: Assessing Scientific, Ethical, Religious, and Policy Issues* (Washington, DC: AAAS Publications Services, 2000), 41, www.aaas.org/spp/sfrl/projects/germline/report.pdf.

11. Jonathan Glover, *What Sort of People Should There Be?* (New York: Penguin, 1984), 149 (quoting John Mackie).

12. Gregory Stock, *Redesigning Humans: Our Inevitable Genetic Future* (New York: Houghton Mifflin, 2002), 115.

13. Claudia Dreifus, "An Economist Examines the Business of Fertility," *New York Times*, Feb. 28, 2006, F5.

14. Rita Rubin, "Giving Growth a Synthetic Hand: Use of Hormone Sparks Debate," *Dallas Morning News*, July 7, 1986, A1.

15. Lori Andrews, *The Clone Age: Adventures in the New World of Reproductive Technology* (New York: Henry Holt, 1999), 142–44.

16. Glover, *What Sort of People?*, 54.

17. Bill McKibben, *Enough: Staying Human in an Engineered Age* (New York: Owl Books, 2004), 209.

18. Michael Sandel, "The Case against Perfection," *Atlantic Monthly*, Apr. 2004, 54.

19. Leon Kass, "The Wisdom of Repugnance," *New Republic* 216, no. 22 (1997): 17–26.

20. Leon Kass, "Triumph or Tragedy? The Moral Meaning of Genetic Technology," *American Journal of Jurisprudence* 45 (2000): 1–16.

21. Council for Responsible Genetics, "Human Germ Line Manipulation."

22. Mary Jo Feldstein, "The Cost of Conception," *St. Louis Post-Dispatch*, Aug. 6, 2006, A1.

23. U.S. Census Bureau, "Income, Earnings, and Poverty Data from the 2007 American Community Survey," www.census.gov/prod/2008pubs/acs-09.pdf.

24. Jan Hoffman, "Here's Looking at Me, Kid," *New York Times*, July 20, 2008, ST1.

25. George J. Annas, "The Man on the Moon, Immortality, and Other Millennial Myths: The Prospects and Perils of Human Genetic Engineering," *Emory Law Journal* 49 (2000): 753, 770.

26. Jonathan Glover, *Choosing Children: Genes, Disability, and Design* (Oxford: Oxford University Press, 2006), 72.

27. Leon R. Kass, *Life, Liberty, and the Defense of Dignity* (San Francisco: Encounter Books, 2002), 131.

28. The President's Council on Bioethics, "Beyond Therapy: Biotechnology and the Pursuit of Perfection" (2003), 37, http://bioethics.georgetown.edu/ pcbe/reports/beyondtherapy/beyond_therapy_final_webcorrected.pdf.

29. Frankel and Chapman, *Human Inheritable Genetic Modifications*, 39.

30. Harnicher v. Utah Medical Center, 962 P.2d 67 (Utah 1998).

31. Joel Feinberg, "Autonomy, Sovereignty, and Privacy: Moral Ideals in the Constitution?," *Notre Dame Law Review* 58 (1983): 445–92.

32. Julian Savulescu, "Procreative Beneficence: Why We Should Select the Best Children," *Bioethics* 15, no. 5–6 (2001): 413–26.

33. Glover, *Choosing Children*, 74.

34. Nick Bostrom, "In Defence of Posthuman Dignity," *Bioethics* 19, no. 3 (2005): 202–14.

35. Erynn S. Gordon et al., "Nondisease Genetic Testing: Reporting of Muscle SNPs Shows Effects on Self-Concept and Health Orientation Scales," *European Journal of Human Genetics* 13 (2005): 1047–54.

36. Kass, "Wisdom of Repugnance."

37. Yuri Levin, "The Paradox of Conservative Bioethics," *New Atlantis* 1 (2003): 53–65.

38. Gregory E. Kaebnick, "Reasons of the Heart: Emotion, Rationality, and the 'Wisdom of Repugnance,'" *Hastings Center Report* 38, no. 4 (2008): 36–45.

39. Julian Savulescu, "Human-Animal Transgenesis and Chimeras Might Be an Expression of Our Humanity," *American Journal of Bioethics* 3 (2003): 22–25.

40. "NIH Stem Cell Information Basics," http://stemcells.nih.gov/info/basics/ basics6.asp.

41. D. S. An et al., "Use of a Novel Chimeric Mouse Model with a Functionally Active Human Immune System to Study Human Immunodeficiency Virus Type 1 Infection," *Clinical and Vaccine Immunology* 14, no. 4 (2007): 391–96.

42. Rick Weiss, "Human Brain Cells Are Grown in Mice: Success Is Encouraging for Stem Cell Therapies," *Washington Post*, Dec. 13, 2005, www.washington post.com/wp-dyn/content/article/2005/12/12/AR2005121201388.html.

43. Center for Humanized SCID Mouse Models, SUNY Upstate Medical Center, www.upstate.edu/microb/scidcenter/.

44. Mark Henderson, "Transplant of Pig Organs into Humans Closer after Stem Cell Breakthrough," *Sunday Times*, June 3, 2009, www.timesonline.co.uk/ tol/life_and_style/health/article6417499.ece.

45. "Sheep Chimera Makes Human Blood Cells," www.thefreelibrary.com/ Sheep chimera makes human blood cells.-a012079903.

46. Jason Scott Robert and Françoise Baylis, "Crossing Species Boundaries," *American Journal of Bioethics* 3, no. 3 (2003): 1–13.

47. Václav Ourednik et al., "Segregation of Human Neural Stem Cells in the Developing Primate Forebrain," *Science* 293, no. 5536 (2001): 1820–24.

48. Jamie Shreve, "The Other Stem-Cell Debate," *New York Times Magazine*, Apr. 10, 2005, 42.

49. National Academy of Sciences News Release, "Guidelines Released for Embryonic Stem Cell Research," Apr. 26, 2005, www8.nationalacademies.org/onpinews/newsitem.aspx?RecordID=11278.

50. Rebecca D. Williams, "Organ Transplants from Animals: Examining the Possibilities," U.S. Food and Drug Administration, http://web.archive.org/web/20071210031618/http://www.fda.gov/fdac/features/596_xeno.html.

51. Biotechnology Industry Organization (BIO), "Xenotransplantation: The Benefits and Risks of Special Organ Transplantation," www.bio.org/articles/xeno transplantation-benefits-and-risks-special-organ-transplantation.

52. Williams, "Organ Transplants from Animals."

53. Ruwani Wijeyekoon and Roger A. Barker, "Cell Replacement Therapy for Parkinson's Disease," *Biochimica et Biophysica Acta* 1792 (2009): 688–702.

54. BIO, "Xenotransplantation."

55. Ibid., 11.

56. Andrea L. Bonnicksen, *Chimeras, Hybrids, and Interspecies Research: Politics and Policymaking* (Washington, DC: Georgetown University Press, 2009), 10.

57. Glover, *What Sort of People?*, 40–41.

58. Ibid., 38.

59. Leon Kass, "Ageless Bodies, Happy Souls: Biotechnology and the Pursuit of Perfection," *New Atlantis* 1 (2003): 9–28.

60. Glover, *What Sort of People?*, 55.

61. Frances Fukuyama, *Our Posthuman Future: Consequences of the Biotechnology Revolution* (New York: Farrar, Strauss and Giroux, 2002), 160.

62. Annas, "Man on the Moon," 753, 771–72.

63. Council of Europe, "Convention for the Protection of Human Rights and Dignity of the Human Being with regard to the Application of Biology and Medicine: Convention on Human Rights and Biomedicine," art. 13 (ETS no. 164), http://conventions.coe.int/treaty/en/Reports/Html/164.htm.

64. Council of Europe, "Convention," art. 2.

65. Bostrom, "In Defence of Posthuman Dignity."

66. Fukuyama, *Our Posthuman Future*, 149.

67. Alex A. S. Weir, Jackie Chappell, and Alex Kacelnik, "Shaping of Hooks in New Caledonian Crows," *Science* 297, no. 5583 (2002): 981.

68. Karim Ouattara, Alban Lemasson, and Klaus Zuberbühler, "Campbell's Monkeys Concatenate Vocalizations into Context-Specific Call Sequences," *Proceedings of the National Academy of Sciences* 106, no. 51 (2009): 22026–31.

69. Bo Gräslund, *Early Humans and Their World* (London: Routledge, 2005), 7.

70. S. F. Brosnan and F. B. M. de Waal, "Monkeys Reject Unequal Pay," *Nature* 425 (2003): 297–99.

71. Dean Falk, *Braindance* (Gainesville: University Press of Florida, 2004), 188.

72. Pamela R. Willoughby, *The Evolution of Modern Humans in Africa: A Comprehensive Guide* (Lanham, MD: AltaMira Press, 2007), 322.

73. Susannah Baruch et al., "Human Germline Genetic Modifications: Issues and Options for Policymakers" (Genetics and Public Policy Center, 2005), 33.

74. Nicholas Wade, "Scientist at Work: Leon R. Kass; Moralist of Science Ponders Its Power," *New York Times*, Mar. 19, 2002, www.nytimes.com/2002/03/19/science/scientist-at-work-leon-r-kass-moralist-of-science-ponders-its-power.html.

75. E. Joshua Rosenkranz, "Custom Kids and the Moral Duty to Genetically Engineer Our Children," *High Technology Journal* 2, no. 1 (1987): 3–53.

76. Joseph Fletcher, "Ethical Aspects of Genetic Controls," NEJM 1971, quoted in Robert T. Francoeur, "We Can—We Must: Reflections on the Technological Imperative," *Journal of Theological Studies* 33 (1972): 428–39.

77. Patricia McDougall et al., "The Consequences of Childhood Peer Rejection, in *Interpersonal Rejection*, ed. Mark R. Leary (New York: Oxford University Press, 2001), 213–47.

78. Bostrom, "In Defence of Posthuman Dignity."

79. "The World's First Test Tube Baby," PBS, www.pbs.org/wgbh/american experience/features/general-article/babies-worlds-first/.

80. Kate Benson, "A Shower of Babies as IVF Turns 30," *Sydney Morning Herald* (Australia), July 25, 2008, 7.

81. John Harris, *Enhancing Evolution: The Ethical Case for Making Better People* (Princeton, NJ: Princeton University Press, 2007), 38.

82. Ann Hulbert, "The Prodigy Puzzle," *New York Times Sunday Magazine*, Nov. 24, 2005, 64.

CHAPTER FIVE. *Broader Consequences for Society*

1. U.S. Census Bureau, "Income, Earnings, and Poverty Data from the 2007 American Community Survey," www.census.gov/prod/2008pubs/acs-09.pdf.

2. Nancy Kress, *Beggars in Spain* (New York: Avon Books, 1993), 37.

3. Maxwell J. Mehlman, *Wondergenes: Genetic Enhancement and the Future of Society* (Bloomington: Indiana University Press, 2003), 110.

4. David Cay Johnston, "The Gap between Rich and Poor Grows in the United States," *International Herald Tribune*, Mar. 29, 2007, www.iht.com/articles/2007/03/29/business/income.4.php.

5. G. William Domhoff, "Wealth, Income, and Power," sociology.ucsc.edu/whorulesamerica/power/wealth.html.

6. Robert Trigaux, "Wealth," *St. Petersburg Times* (Florida), Oct. 3, 2010, 1P.

7. Gary Young, "In the US, Class War Still Means Just One Thing: The Rich Attacking the Poor," *Guardian*, Sept. 3, 2007, www.guardian.co.uk/commentis free/story/0,,2161252,00.html.

8. Domhoff, "Wealth, Income, and Power."

9. Ibid.

10. Council for Responsible Genetics, "Position Paper on Human Germ Line Manipulation," *Human Therapy* 4 (1992): 35–37.

11. Emily Marden and Dorothy Nelkin, "Displaced Agendas: Current Regulatory Strategies for Germline Gene Therapy," *McGill Law Journal* 45 (2000): 461, 467.

12. Susannah Baruch et al., "Human Germline Genetic Modifications: Issues and Options for Policymakers" (Genetics and Public Policy Center, 2005), 36.

13. George J. Annas, "The Man on the Moon, Immortality, and Other Millennial Myths: The Prospects and Perils of Human Genetic Engineering," *Emory Law Journal* 49 (2000): 753, 772–73.

14. Ian Tattersall, "Once We Were Not Alone," *Scientific American*, Special edition 2003, 21, 25.

15. Claes Ramel, "Man as a Biological Species," *AMBIO* 21, no. 1 (1992): 75–78.

16. David Sloan Wilson, *Evolution for Everyone: How Darwin's Theory Can Change the Way We Think about Our Lives* (New York: Delacort Press, 2007), 218.

17. Stephen R. Palumbi, *The Evolution Explosion: How Humans Cause Rapid Evolutionary Change* (New York: W. W. Norton, 2001), 11–12.

18. Ernst Mayr, *What Evolution Is* (New York: Basic Books, 2001), 195–96.

19. Nicholas Wade, *Before the Dawn* (New York: Penguin, 2006), 134–38.

20. Joseph Fletcher, *The Ethics of Genetic Control* (Buffalo, NY: Prometheus Books, 1988), 19.

21. Mark S. Frankel and Audrey R. Chapman, *Human Inheritable Genetic Modifications: Assessing Scientific, Ethical, Religious, and Policy Issues* (Washington, DC: AAAS Publications Services, 2000), www.aaas.org/spp/sfrl/projects/germline/report.pdf.

22. Jacqueline L. Salmon, "Most Americans Believe in Higher Power, Poll Finds," *Washington Post*, June 24, 2008, A02.

23. Fletcher, *Ethics of Genetic Control*, 200.

24. Robert T. Francoeur, "We Can—We Must: Reflections on the Technological Imperative," *Journal of Theological Studies* 33 (1972): 428–39.

25. Ronald Dworkin, *Sovereign Virtue: The Theory and Practice of Equality* (Cambridge, MA: Harvard University Press, 2002), 446.

26. Edwards v. Aguillard, 482 U.S. 578 (1987).

27. Ibid., 634.

28. Kitzmiller v. Dover Area School District, 400 F. Supp. 2d 707, 708 (M.D. Pa. 2005).

29. Center for Science and Culture, "Definition of Intelligent Design," www.intelligentdesign.org/whatisid.php.

30. *Kitsmiller*, 709.

31. Ibid., 765.

32. Phyllis Schlafly, "False Judge Makes Mockery of Case for 'Intelligent Design,'" http://townhall.com/columnists/PhyllisSchlafly/2006/01/02/false_judge_makes_mockery_of_case_for_intelligent_design/page/1.

33. Claudia Wallis, "The Evolution Wars," *Time*, Aug. 7, 2005, www.time.com/time/magazine/article/0,9171,1090909-1,00.html.

34. La. Senate Bill 733, reg. sess. (2008), www.legis.state.la.us/billdata/streamdocument.asp?did=482728.

35. James Gill, "Mad Scientists," *Times-Picayune* (New Orleans), Feb. 18, 2009, 7.

36. Guy Trebay, "He's Pregnant. You're Speechless," *New York Times*, June 22, 2008, www.nytimes.com/2008/06/22/fashion/22pregnant.html?pagewanted=1&_r=1.

37. *In re* Marriage of Buzzanca, 72 Cal. Rptr. 2d 280 (Cal. App. 1998).

38. National Abortion Federation, "NAF Violence and Disruption Statistics," www.prochoice.org/pubs_research/publications/downloads/about_abortion/stats_table2009.pdf.

39. Charles J. Epstein, "1996 ASHG Presidential Address: Toward the 21st Century," *American Journal of Human Genetics* 60 (1997): 1–9.

40. Peter M. Buston and Stephen T. Emlen, "Cognitive Processes Underlying Human Mate Choice: The Relationship between Self-Perception and Mate Preference in Western Society," *Proceedings of the National Academy of Sciences* 100, no. 15 (2003): 8805–10.

41. Oliver Curry, "The Bravo Evolution Report," www.kingkraal.com/bravo.pdf.

42. Caroline Iggulden, "Good News. All Men Will Have Big Willies. Bad News. It Won't Happen till Year 3000," *Sun* (London), Oct. 17, 2006, 10, http://docs.newsbank.com/s/InfoWeb/aggdocs/AWNB/114D414153E114B0/1625BF23780341F390CF767A4DFF3638?p_multi=LSNB&s_lang=en-US.

43. H. Allen Orr, "Testing Natural Selection," *Scientific American*, Jan. 2009, 44, 50.

44. Mayr, *What Evolution Is*, 261.

45. Jason deParle, "A World on the Move," *New York Times*, June 27, 2010, WK1.

46. Peter Ward, *Future Evolution: An Illuminated History of Life to Come* (New York: Times Books, 2001), 140.

47. Gregory Stock, *Redesigning Humans: Our Inevitable Genetic Future* (New York: Houghton Mifflin, 2002), 182.

48. Nick Bostrom, "In Defence of Posthuman Dignity," *Bioethics* 19, no. 3 (2005): 202–14.

49. Nancy M. Tanner, "Gathering by Females: The Chimpanzee Model Revisited and the Gathering Hypothesis, in *The Evolution of Human Behavior Primate Models*, ed. W. G. Kinzey (Albany: SUNY Press, 1987), 3, 14, 18.

50. Dean Falk, *Braindance* (Gainesville: University Press of Florida, 2004), 94.

51. A. C. Perdeck, "The Isolating Value of Specific Song Patterns in Two Sibling Species of Grasshoppers (*Chorthippus brunneus* Thunb. and *C. biguttulus* L.)," *Behaviour* 12, no. 1/2 (1958): 1–75.

52. Ward, *Future Evolution*, 148.

53. Nicholas Wade, *Before the Dawn* (New York: Penguin, 2006), 28, fig. 2.3.

54. Robert Foley, *Unknown Boundaries* (Cambridge: Cambridge University Press, 2005), 17; Kate Wong, "Who Were the Neanderthals?," *Scientific American*, Special edition 2003, 28, 36.

55. Richard E. Green et al., "A Draft Sequence of the Neandertal Genome," *Science* 328, no. 5979 (2010): 710–22.

56. Ibid.

57. Nicholas Wade, "Neanderthal Women Joined Men in the Hunt," *New York Times*, Dec. 5, 2006, www.nytimes.com/2006/12/05/science/05nean.html.

58. Marcia S. Ponce de Leon, "Neanderthal Brain Size at Birth Provides Insights into the Evolution of Human Life History," *Proceedings of the National Academy of Sciences* 105, no. 37 (2008): 13764–68.

59. Wong, "Who Were the Neanderthals?," 36.

60. Thomas Wynn and Frederick Coolidge, "Why Not Cognition?," *Current Anthropology* 49, no. 5 (2008): 895–97.

61. Wong, "Who Were the Neanderthals?," 34–35.

62. Clive Finlayson and José S. Carrión, "Rapid Ecological Turnover and Its Impact on Neanderthal and Other Human Populations," *Trends in Ecology and Evolution* 22, no. 4 (2007): 213–22.

63. P. C. Tzedakis, "Seven Ambiguities in the Mediterranean Palaeoenvironmental Narrative," *Quaternary Science Reviews* 26 (2007): 2042–66.

64. Kristian J. Herrera et al., "To What Extent Did Neanderthals and Modern Humans Interact?," *Biological Reviews* 84 (2009): 245–57.

65. Eduardo Moreno, "A War-Prone Tribe Migrated out of Africa to Populate the World," *Nature Precedings*, Mar. 22, 2010, http://precedings.nature.com/documents/4303/version/1.

66. Robin McKie, "How Neanderthals Met a Grisly Fate: Devoured by Humans," *Guardian*, May 17, 2009, www.guardian.co.uk/science/2009/may/17/neanderthals-cannibalism-anthropological-sciences-journal.

67. It is possible, though, that the inability to interbreed would make the relationship between *Homo sapiens* and *Homo mutatus* less hostile, since they would not be competing for the same mates.

68. Annas, "Man on the Moon," 753, 773.

69. Mayr, *What Evolution Is*, 125.

70. Bostrom, "In Defence of Posthuman Dignity."

71. Richard Dawkins, "God's Utility Function," *Scientific American*, Nov. 1995, 85.

72. Jonathan Glover, *What Sort of People Should There Be?* (Hammondsworth, UK: Penguin Books, 1984), 179.

CHAPTER SIX. *The End of the Human Lineage*

1. A. Purvis et al., "Predicting Extinction Risk in Declining Species," *Proceedings of the Royal Society B: Biological Sciences* 267, no. 1456 (2000): 1947–52;

P. L. Munday, "Habitat Loss, Resource Specialization, and Extinction on Coral Reefs," *Global Change Biology* 10, no. 10 (2004): 1642–47.

2. U.S. Centers for Disease Control and Prevention, "DES Update," www.cdc .gov/des/consumers/about/effects.html.

3. John A. McLachlan, "Commentary: Prenatal Exposure to Diethylstilbestrol (DES): A Continuing Story," *International Journal of Epidemiology* 35, no. 4 (2006): 868–70.

4. Council for Responsible Genetics, "Position Paper on Human Germ Line Manipulation," *Human Therapy* 4 (1992): 35, 37.

5. Chandra Wickramasinghe, Milton Wainright, and Jayant Narlikar, "SARS—a Clue to Its Origins?," *Lancet* 361, no. 9371 (2003): 1832.

6. Kurt J. Leonard, "Black Stem Rust Biology and Threat to Wheat Growers" (U.S. Dept. of Agriculture, Agricultural Research Service, 2001), www.ars.usda .gov/Main/docs.htm?docid=10755.

7. A. J. Ullstrup, "The Impacts of the Southern Corn Leaf Blight Epidemics of 1970–1971," *Annual Review of Phytopathology* 10 (1972): 37–50.

8. Catharina Japikse, "The Irish Potato Famine," *EPA Journal* 20, no. 3/4 (1994): 44.

9. Cecil Woodham-Smith, *The Great Hunger* (London: Hamish Hamilton, 1962), 35.

10. James Meek, "Yes—in 10 Years We May Have No Bananas," *Guardian Weekly*, Jan. 16, 2003, www.guardian.co.uk/science/2003/jan/16/gm.science.

11. Jose Esquinas-Alcazar, "Protecting Crop Genetic Diversity for Food Security: Political, Ethical, and Technical Challenges," *Nature Reviews Genetics* 6 (2005): 946–53.

12. American Society for Reproductive Medicine, "Minimal Genetic Screening for Gamete Donors," *Fertility and Sterility* 70 (1998): 12–13.

13. American Academy of Pediatrics, "Report of the Task Force on Newborn Screening, Published as a Call for a National Agenda on State Newborn Screening Programs," *Pediatrics* 106, suppl. 2 (2000): 389.

14. American Society of Human Genetics and American College of Medical Genetics, "Points to Consider: Ethical, Legal, and Psychosocial Implications of Genetic Testing in Children and Adolescents," *American Journal of Human Genetics* 57, no. 1233 (1995); Council on Ethical and Judicial Affairs, American Medical Association, Current Opinion 2.138, "Code of Medical Ethics" (1995).

15. American Medical Association Council on Ethical and Judicial Affairs, Current Opinion E-2.11, "Gene Therapy" (updated 1994).

16. American Medical Association Council on Ethical and Judicial Affairs, "Report on Ethical Issues Related to Prenatal Genetic Tests," *Archives of Family Medicine* 3 (1994): 633, 637–39.

17. National Advisory Council for Human Genome Research, "Statement on the Use of DNA Testing for Presymptomatic Identification of Cancer Risk," *Journal of the American Medical Association* 10 (1997): 159–66.

18. American Society of Clinical Oncology, "Genetic Testing for Cancer Susceptibility," *Journal of Clinical Oncology* 14 (1996): 1730–36.

19. Marc-Andre Gagnon and Joel Lexchin, "The Cost of Pushing Pills: A New Estimate of Pharmaceutical Promotion Expenditures in the United States," *PLoS Medicine* 5, no. 1 (2008): e1, doi:10.1371/journal.pmed.0050001.

20. Julie M. Donohue, Marisa Cevasco, and Meredith B. Rosenthal, "A Decade of Direct-to-Consumer Advertising of Prescription Drugs," *New England Journal of Medicine* 357, no. 7 (2007): 673–81.

21. Daniel Callahan, "Too Much of a Good Thing: How Splendid Technologies Can Go Wrong," *Hastings Center Report* 33, no. 2 (2003): 19.

22. SEC v. Citigroup Global Markets, SEC 2003, www.sec.gov/litigation/complaints/compl8111.htm.

23. Julian Savulescu, "Procreative Beneficence: Why We Should Select the Best Children," *Bioethics* 15, no. 5–6 (2001): 413–26.

24. Mark S. Frankel and Audrey R. Chapman, *Human Inheritable Genetic Modifications: Assessing Scientific, Ethical, Religious, and Policy Issues* (Washington, DC: AAAS Publications Services, 2000), 35, www.aaas.org/spp/sfrl/projects/germline/report.pdf.

25. Martin Bobrow, "Redrafted Chinese Law Remains Eugenic," *Journal of Medical Genetics* 32 (1995): 409.

26. *Time Asia*, "The Creation of Yao Ming," adapted from *Operation Yao Ming* by Brook Lamar (New York: Penguin, 2005), www.time.com/time/asia/covers/501051114/story.html.

27. National Research Council, Board on Army Science and Technology, Committee on Opportunities in Biotechnology for Future Army Applications, *Opportunities in Biotechnology for Future Army Applications* (Washington, DC: National Academies Press, 2001), 64.

28. Ashley R. Melson, "Bioterrorism, Biodefense, and Biotechnology in the Military: A Comparative Analysis of Legal and Ethical Issues in the Research, Development, and Use of Biotechnological Products on American and British Soldiers," *Albany Law Journal of Science & Technology* 14 (2004): 497–535n41.

29. JASON 2010, *The $100 Genome: Implications for DoD* (Report No. JSR-10-100, Dec. 2010), www.fas.org/irp/agency/dod/jason/hundred.pdf.

30. Jonathan D. Moreno, *Mind Wars: Brain Research and National Defense* (Washington, DC: Dana Press, 2006).

31. Francis Galton, "Eugenics: Its Definition, Scope, and Aims," *American Journal of Sociology* 10, no. 1 (1904): 1.

32. Dan Balz, "Sweden Sterilized Thousands of 'Useless' Citizens for Decades," *Washington Post*, Aug. 29, 1997, A1.

33. Nick Bostrom, "In Defence of Posthuman Dignity," *Bioethics* 19, no. 3 (2005): 202–14.

34. Allen Buchanan et al., *From Chance to Choice: Genetics and Justice* (Cambridge: Cambridge University Press, 2000), 28.

35. Bostrom, "In Defence of Posthuman Dignity," 206.

36. Maxwell J. Mehlman, "Modern Eugenics and the Law," in *A Century of Eugenics in America: From the "Indiana Experiment" to the Human Genome Era*, ed. Paul A. Lombardo (Bloomington: Indiana University Press, 2010).

37. "Kids Swap DNA for Fairground Rides," *Scientific American*, Sept. 1, 2010, www.scientificamerican.com/article.cfm?id=kids-swap-dna-for-fairground.

38. Duane Alexander and Peter C. van Dyck, "A Vision of the Future of Newborn Screening," *Pediatrics* 117 (2006): S350, S352.

39. Jon W. Gordon, "Genetic Enhancement in Humans," *Science* 283 (1999): 2023.

40. Robert Nozick, *Anarchy, State and Utopia* (New York: Basic Books, 1974), 315.

41. Guido Barbujani and Vincenza Colonna, "Human Genome Diversity: Frequently Asked Questions," *Trends in Genetics*, May 13, 2010, www.sciencedirect.com/science/article/pii/S0168952510000788.

42. Jonathan Glover, *What Sort of People Should There Be?* (Hammondsworth, UK: Penguin Books, 1984), 48.

43. "Illumina Announces New Pricing for Its Individual Genome Sequencing Service," http://investor.illumina.com/phoenix.zhtml?c=121127&p=irol-newsArticle&ID=1434418&highlight=.

44. National Institutes of Health, National Human Genome Research Institute, "NHGRI Uses Recovery Act Funds to Accelerate Genome Research to Improve Human Health," www.genome.gov/27534233.

45. JASON, *$100 Genome*.

46. Peter Ward, *Future Evolution: An Illuminated History of Life to Come* (New York: Times Books, 2001), 147.

47. Julianna Kettlewell, "'Frozen Ark' to Save Animal DNA," *BBC Online*, July 27, 2004, http://news.bbc.co.uk/2/hi/science/nature/3928411.stm.

48. Didier Sornette, "Predictability of Catastrophic Events: Material Rupture, Earthquakes, Turbulence, Financial Crashes, and Human Birth," *Proceedings of the National Academy of Sciences* 99, suppl. 1 (2002): 2522–29.

49. U.S. Geologic Survey CoreCast, May 20, 2009, www.usgs.gov/corecast/details.asp?ID=76.

50. Peter Tyson, "Forecasting Volcanic Eruptions," *NOVA*, PBS, June 2005, www.pbs.org/wgbh/nova/volcanocity/boom.html.

51. Bureau of Economic Analysis, "National Economic Accounts: Gross Domestic Product," www.bea.gov/national/index.htm#gdp.

52. Editorial, "Will Cloning Beget Disasters?," *Wall Street Journal*, May 2, 1997, A14.

53. Purvis et al., "Predicting Extinction Risk."

54. Martin Walker, "The World's New Numbers," *Wilson Quarterly*, Spring 2009, www.wilsonquarterly.com/article.cfm?aid=1408.

55. Marcel Cardillo and Adrian Lister, "Death in the Slow Lane," *Nature* 419, no. 6906 (2002): 440.

56. Nicholas Wade, *Before the Dawn* (New York: Penguin Press, 2006), 24.

57. Lei Jin et al., "Preliminary Evidence regarding the Hypothesis That the Sex Ratio at Sexual Maturity May Affect Longevity in Men," *Demography* 47, no. 3 (2010): 579–86.

58. Maureen J. Graham, Ulla Larsen, and Xiping Xu, "Son Preference in Anhui Province, China," *International Family Planning Perspectives* 24, no. 2 (1998), www.guttmacher.org/pubs/journals/2407298.html.

59. Zou Hanru, "Unbalanced Ratio Caused by Rural Woes," *China Daily*, Oct. 13, 2006, www.chinadaily.com.cn/opinion/2006-10/13/content_707335.htm.

60. Walker, "World's New Numbers."

61. Valerie M. Hudson and Andrea Den Boer, "A Surplus of Men, A Deficit of Peace: Security and Sex Ratios in Asia's Largest States," *International Security* 26, no. 4 (2002): 5–38.

62. Rohini Pande and Anju Malhotra, "Son Preference and Daughter Neglect in India: What Happens to Living Girls?," International Center for Research on Women 2006, www.icrw.org/publications/son-preference-and-daughter-neglect -india.

63. Janice Shaw Crouse, "United States Resolution Shanghaied by China and India," Concerned Women for America, Mar. 9, 2007, www.cwfa.org/articledis play.asp?id=12532&department=BLI&categoryid=reports&subcategoryid=bliun.

64. GeoHive, www.xist.org/earth/pop_gender.aspx.

65. Genetics and Public Policy Center, "Survey of Fertility Clinics: Selection Technologies Widespread in the U.S.," *Genetic Crossroads*, Oct. 20, 2006, www .geneticsandsociety.org/article.php?id=2587.

66. Lori B. Andrews, *The Clone Age: Adventures in the New World of Reproductive Technology* (New York: Henry Holt, 2001), 142–44.

67. Ernst Mayr, *What Evolution Is* (New York: Basic Books, 2001), 140–44.

68. Julian Savulescu, "Human-Animal Transgenesis and Chimeras Might Be an Expression of Our Humanity," *American Journal of Bioethics* 3, no. 3 (2003): 22–25.

69. James Watson, "Engineering the Human Germline Symposium," Summary Report 1998, www.ess.ucla.edu/huge/report.html, quoted in Emily Marden and Dorothy Nelkin, "Displaced Agendas: Current Regulatory Strategies for Germline Gene Therapy," *McGill Law Journal* 45 (2000): 461, 468.

CHAPTER SEVEN. *Evolution by Nature or by Human Design?*

1. Leon Kass, "The Moral Meaning of Genetic Technology," *Commentary*, Sept. 1999, 32, 38.

2. G. A. Harrison et al., *Human Biology*, 3rd ed. (Oxford: Oxford University Press, 1988), 214–15.

3. H. Allen Orr, "Testing Natural Selection," *Scientific American*, Jan. 2009, 44, 50.

4. Ernst Mayr, "The Objects of Selection," *Proceedings of the National Academy of Sciences* 94, no. 6 (1997): 2091–94.

5. Ernst Mayr, *What Evolution Is* (New York: Basic Books, 2001), 135.

6. Dean Falk, *Braindance* (Gainesville: University Press of Florida, 2004), 3.

7. Joseph Fletcher, *The Ethics of Genetic Control* (Buffalo, NY: Prometheus Books, 1988), 296.

8. Mayr, *What Evolution Is*, 121.

9. Nicholas Wade, *Before the Dawn* (New York: Penguin, 2006), 177.

10. Mayr, *What Evolution Is*, 278.

11. Stephen J. Gould, "The Evolution of Life on Earth," in *The Richness of Life: The Essential Stephen J. Gould*, ed. Steven Rose (New York: W. W. Norton, 2006), 216–17.

12. Ian Tattersall, "Once We Were Not Alone," *Scientific American*, Special edition 2003, 21, 24.

13. Ibid., 109.

14. Gould, "Evolution of Life," 215.

15. *Bill Moyers: Genesis, A Living Conversation*, "In God's Image," PBS, www.pbs.org/wnet/genesis/program2.html.

16. Gould, "Evolution of Life," 214.

17. Ibid., 285.

18. Ann Gibbons, *The First Human: The Race to Discover Our Earliest Ancestors* (New York: First Anchor Books, 2007), 166.

19. John Lichfield and Declan Walsh, "Fossil Men Feud over Our 'Oldest Ancestor,'" *Independent* (UK), Feb. 18, 2001.

20. Donald Johanson, "Lucy's Story," Institute of Human Origins, Arizona State University, http://iho.asu.edu/lucy.html#name.

21. Ann Gibbons, "Breakthrough of the Year: *Ardipithecus ramidus*," *Science* 18 (2009): 1598–99.

22. Bo Gräslund, *Early Humans and Their World* (London: Routledge, 2005), 65.

23. Falk, *Braindance*, 96.

24. Ibid.

25. Ibid.

26. In fact, Kent State University anthropologist Owen Lovejoy, who incidentally was the researcher who analyzed Ardi's bones below the neck (Jamie Shreeve, "Oldest Skeleton of Human Ancestor Found," *National Geographic News*, Oct. 1, 2009, http://news.nationalgeographic.com/news/2009/10/091001-oldest-human-skeleton-ardi-missing-link-chimps-ardipithecus-ramidus.html), claims that bipedalism was selected for not because of the savannah, but because of changes in sexual behavior, mainly monogamy, which favored males who could more easily carry provisions and other goods home to woo wives and supply their families. C. Owen Lovejoy, "The Origin of Man," *Science* 211, no. 4480 (1981): 341–50.

27. Gräslund, *Early Humans*, 68.

28. Ibid., 67.

29. Ibid., 65.

30. Falk, *Braindance*, 165–66.

31. Ibid., 90–91.

32. Ibid.

33. Ibid., 65–66.

34. Ibid., 88.

35. Mayr, *What Evolution Is*, 282.

36. Falk, *Braindance*, 4–5.

37. Rex Dalton, "Fossil Primate Challenges Ida's Place," *Nature* 461, no. 7267 (2009): 1040.

38. Ibid.

39. Erik R. Seiffert et al., "Convergent Evolution of Anthropoid-Like Adaptations in Eocene Adapiform Primates," *Nature* 461, no. 7267 (2009): 1027–1162.

40. Kate Wong, "Rethinking the Hobbits of Indonesia," *Scientific American* 301, no. 5 (2009): 72.

41. Falk, *Braindance*, 4.

42. Pamela R. Willoughby, *The Evolution of Modern Humans in Africa: A Comprehensive Guide* (Lanham, MD: AltaMira Press, 2007), 319.

43. Falk, *Braindance*, 4.

44. Willoughby, *Evolution of Modern Humans*, 321.

45. Alan G. Thorne and Milford H. Wolpoff, "The Multiregional Evolution of Humans," *Scientific American*, Special edition 2003, 46, 48.

46. Rebecca L. Cann and Allan C. Wilson, "The Recent African Genesis of Humans," *Scientific American*, Special edition 2003, 54, 56.

47. Ibid., 55.

48. Rebecca L. Cann, Mark Stoneking, and Allan C. Wilson, "Mitochondrial DNA and Human Evolution," *Nature* 325 (1987): 31–36.

49. Cann and Wilson, "Recent African Genesis," 54, 55.

50. Ibid.

51. Ibid.

52. Wolpoff and Thorne explain what they mean with the following illustration:

> Imagine an immigrant neighborhood in a large city where all the families share a surname. An observer might assume that all these families were descended from a single successful immigrant family that completely replaced its neighbors. An alternative explanation is that many families immigrated to the neighborhood and intermarried; over time, all the surnames but one were randomly eliminated through the occasional appearance of families that had no sons to carry on their names. The surviving family name would have come from a single immigrant, but all the immigrants would have contributed to the genes of the modern population. In the same way, generations without daughters could have extinguished some lines of mitochondrial DNA from Eve's descendants and her contemporaries.

Thorne and Wolpoff, "Multiregional Evolution of Humans," 51.

53. Ibid., 61.

54. Sean B. Carroll, Benjamin Prud'homme, and Nicholas Gompel, "Regulating Evolution," *Scientific American*, May 2008, 62.

55. S. B. Carroll, "The Origins of Form," *Natural History* 114, no. 9 (2005): 58–63.

56. Leon Kass, "Ageless Bodies, Happy Souls: Biotechnology and the Pursuit of Perfection," *New Atlantis* 1 (2003): 9–28.

57. Mayr, *What Evolution Is*, 252.

58. Falk, *Braindance*, 6.

59. Carroll, Prud'homme, and Gompel, "Regulating Evolution," 62.

CHAPTER EIGHT. *Protecting the Children*

1. Rosario M. Isasi, Thu Minh Nguyen, and Bartha M. Knoppers, "National Regulatory Frameworks regarding Human Genetic Modification Technologies (Somatic and Germline Modification)" (Genetics and Public Policy Center, 2006), www.dnapolicy.org/pdf/geneticModification.pdf.

2. Council of Europe, Convention for the Protection of Human Rights and Dignity of the Human Being with regard to the Application of Biology and Medicine: Convention on Human Rights and Biomedicine, Oviedo, 4.IV.1997, http://conventions.coe.int/treaty/en/treaties/html/164.htm.

3. Andrea L. Bonnicksen, *Chimeras, Hybrids, and Interspecies Research: Politics and Policymaking* (Washington, DC: Georgetown University Press, 2009), 3–4, 51, 53–54.

4. Transcript of President Bush's State of the Union Address, *Washington Post*, Jan. 31, 2006, www.washingtonpost.com/wp-dyn/content/article/2006/01/31/AR2006013101468.html.

5. United Kingdom, Human Fertilization and Embryology Act Amendments (2008), www.opsi.gov.uk/acts/acts2008/ukpga_20080022_en_2#pt1-pb1-l1g1.

6. Ibid., Section 4A(3).

7. Alok Jha, "First British Human-Animal Hybrid Embryos Created by Scientists," *Guardian*, Apr. 2, 2008, www.guardian.co.uk/science/2008/apr/02/medicalresearch.ethicsofscience.

8. Murray Wardrop, "Human-Animal Clone Research Halted amid Funding Drought," *Telegraph*, Jan. 13, 2009, www.telegraph.co.uk/science/science-news/4225636/Human-animal-clone-research-halted-amid-funding-drought.html.

9. Steve Connor, "Funding Halter for Stem Cell Research," *Independent* (UK), Jan. 13, 2009, www.independent.co.uk/news/science/funding-halted-for-stem-cell-research-1332000.html.

10. U.S. Food and Drug Administration, Adverse Events Reporting System, www.fda.gov/Drugs/GuidanceComplianceRegulatoryInformation/Surveillance/AdverseDrugEffects/ucm070461.htm.

11. Mark S. Frankel and Audrey R. Chapman, *Human Inheritable Genetic Modifications: Assessing Scientific, Ethical, Religious, and Policy Issues* (Washing-

ton, DC: AAAS Publications Services, 2000), 26, www.aaas.org/spp/sfrl/projects/germline/report.pdf.

12. Ibid., 13.

13. J. Rawls, *A Theory of Justice* (Cambridge, MA: Belknap Press of Harvard University Press, 1971).

14. N. Daniels, *Just Health Care: Studies in Philosophy and Health Policy* (Cambridge: Cambridge University Press, 1985).

15. D. W. Brock and N. Daniels, "Ethical Foundations of the Clinton Administration's Proposed Health Care System," *Journal of the American Medical Association* 271, no. 15 (1994): 1189–96.

16. Wardrop, "Human-Animal Clone Research."

17. MediCal News, "Interactions between Human and Microbial Cells Determine Health, Physical Well-Being: Researchers," June 11, 2010, www.news-medical.net/news/20100611/Interactions-between-human-and-microbial-cells-determine-health-physical-well-being-Researchers.aspx.

18. National Institutes of Health, Human Microbiome Project, http://nihroadmap.nih.gov/hmp/.

19. "Inhuman Genomes: Every Genome on the Planet Is Now Up for Grabs, Including Those That Do Not Yet Exist," *Economist*, June 17, 2010, www.economist.com/node/16349380.

20. Junjie Qin et al., "A Human Gut Microbial Gene Catalogue Established by Metagenomic Sequencing," *Nature* 464 (2009): 59–65.

21. JASON 2010, *The $100 Genome: Implications for DoD* (Report No. JSR-10-100, Dec. 2010), www.fas.org/irp/agency/dod/jason/hundred.pdf.

22. S. R. Gill et al., "Metagenomic Analysis of the Human Distal Gut Microbiome," *Science* 2, no. 312 (2006): 1355–59.

23. Noah Schachtman, "Pig Manure Key to Soldier Chow?," *Wired*, Mar. 27, 2007, www.wired.com/dangerroom/2007/03/darpa_the_penta/#ixzz12OG5D1OD.

24. David J. Griffiths, "Endogenous Retroviruses in the Human Genome Sequence," *Genome Biology* 2, no. 6 (2001): Reviews, http://genomebiology.com/2001/2/6/reviews/1017.

25. Derek Parfit, *Reasons and Persons*, rev. ed. (Oxford: Oxford University Press, 1987), 351–79.

26. John Robertson, "Procreative Liberty and Harm to Offspring in Assisted Reproduction," *American Journal of Law & Medicine* 30 (2004): 7–40.

27. Ronald Dworkin, *Sovereign Virtue: The Theory and Practice of Equality* (Cambridge, MA: Harvard University Press, 2002), 433.

28. Curlender v. Bio-Science Laboratories, 165 Ca, Rptr. 477 (Cal. App. 1980).

29. Doolan v. IVF America (MA) Inc., 2000 WL 33170944 (Mass. Super. 2000).

30. Turpin v. Sortini, 182 Cal. Rptr. 337 (1982).

31. Goldberg v. Ruskin, 499 N.E. 2d 406, 411 (Ill. 1986).

32. Ibid.

33. *Curlender*, 488.

34. *In re* Philip B., 92 Cal App. 3d 796 (Cal. Ct. App. 1979).

35. Newmark v. Williams, 588 A.2d 1108 (Del. 1991).

36. In the matter of Lyle A., 830 NYS2d 486 (Fam. Ct. N.Y. 2006).

37. Hart v. Brown, 289 A.2d 386 (Conn. Sup. Ct. 1972).

38. Alicia Ouellette, "Shaping Parental Authority over Children's Bodies," *Indiana Law Journal* 85 (2010): 955–1002.

39. Ibid.

40. Barbara Bennett Woodhouse, "'Who Owns the Child?': *Meyer* and *Pierce* and the Child as Property," *William and Mary Law Review* 33 (1992): 995–1122.

41. Michael Grossberg, *Governing the Hearth: Law and the Family in Nineteenth-Century America* (Chapel Hill: University of North Carolina Press, 1985), 25.

42. Woodhouse, "Who Owns the Child?," 1035.

43. Ibid., 1046.

44. Steven Mintz and Susan Kellogg, *Domestic Revolutions: A Social History of American Family Life* (New York: Free Press, 1988), 54.

45. Parham v. J.R., 442 U.S. 584, 602 (1979).

46. Ronald M. Green, "Prenatal Autonomy and the Obligation Not to Harm One's Child Genetically," *Journal of Law, Medicine & Ethics* 25, no. 1 (1997): 5–15.

47. Matter of E.J., 465 A.2d 374 (DC Ct. App. 1983).

48. DeMatteo v. DeMatteo, 194 Misc. 2d 640 (Sup. Ct. NY 2002).

49. *In re* Petra, 216 Cal. App. 3d 1163 (1989).

50. *Parham*, 584, 603.

51. Woodhouse, "Who Owns the Child?," 1048.

52. 42 USCA § 5106g(2).

53. Ohio Rev. Code § 2919.22.

54. Laurence B. McCullough and Frank A. Chervenak, "A Critical Analysis of the Concept and Discourse of 'Unborn Child,'" *American Journal of Bioethics* 8 (2008): 34, 36, 38.

55. 18 USCS § 1841(a)(1).

56. Jessica Berg, "Of Elephants and *Embryos:* A Proposed Framework for Legal Personhood," *Hastings Law Journal* 59 (2007): 369–406.

57. Whitner v. State, 492 S.E. 2d 777 (S.C. 1997).

58. AMA Council on Ethical and Judicial Affairs, "Ethical Issues Related to Prenatal Genetic Testing," *Archives of Family Medicine* 3 (1994): 633.

59. Susannah Baruch et al., "Human Germline Genetic Modifications: Issues and Options for Policymakers" (Genetics and Public Policy Center, 2005), 22.

60. Lisa L. Bakhos et al., "Emergency Department Visits for Concussion in Young Child Athletes," *Pediatrics*, Aug. 30, 2010, http://pediatrics.aappublications.org/cgi/reprint/peds.2009-3101v1.

61. 42 CFR § 405.

62. Bruce Ackerman, *Social Justice in the Liberal State* (New Haven, CT: Yale University Press, 1980).

63. U.K. Human Fertilisation and Embryology Authority, "Sex Selection," www.hfea.gov.uk/pgd-sex-selection.html.

64. Martin Gruber and Tim Studt, "Re-emerging U.S. R&D," *R&D Magazine*, www.rdmag.com/Featured-Articles/2009/12/Policy-and-Industry-Re-Emer ging-U-S-R-D/.

65. Susan Crockin, "A Legal Defense for Compensating Research Egg Donors," *Cell Stem Cell* 6, no. 2 (2010): 99–102.

66. David Tuller, "Payment Offers to Egg Donors Prompt Scrutiny," *New York Times*, May 11, 2010, www.geneticsandsociety.org/article.php?id=5194.

67. *Final Report of the National Academies' Human Embryonic Stem Cell Research Advisory Committee and 2010 Amendments to the National Academies' Guidelines for Human Embryonic Stem Cell Research* (Washington, DC: National Academies Press, 2010), http://books.nap.edu/openbook.php?record_id= 12923&page=27.

68. Tuller, "Payment Offers to Egg Donors"; Crockin, "Legal Defense."

69. *NAS Final Report*, 5.

70. Tuller, "Payment Offers to Egg Donors."

71. "Eggs and Sperm from Stem Cells on the Horizon, Ethics Group Tells AAAS Meeting," www.aaas.org/news/releases/2008/0423hinxton.shtml.

72. National Conference of State Legislatures, "Stem Cell Research Laws and Legislative Activity," www.ncsl.org/programs/health/Genetics/embfet.htm.

73. *In re* Marriage of Dahl & Angle, 194 P.3d 834 (Or. Ct. App. 2008). See also Jessica Berg, "Owning Persons: The Application of Property Theory to Embryos and Fetuses," *Wake Forest Law Review* 40 (2005): 159.

74. *NAS Final Report*, 26.

75. Ibid., Section 4A(3).

76. International Society for Stem Cell Research, "Stem Cell Research Regulations in the European Union," www.isscr.org/scientists/legislative.htm.

77. Dirceu B. Greco, Human Cloning and International Governance Public Hearings, Fifteenth Section of the International Bioethics Comm. (IBC) of Unesco (Oct. 29, 2008), www.unesco.org/new/fileadmin/MULTIMEDIA/HQ/ SHS/pdf/Public-Hearings-Cloning-Greco-CONEP-Brazil.pdf.

78. 45 CFR § 46.204(b).

79. 45 CFR § 46.204(e).

80. 45 CFR § 46.407.

81. Anne Wood, Christine Grady, and Ezekiel J. Emanuel, "Regional Ethics Organizations for Protection of Human Research Participants," *Nature Medicine* 10 (2004): 1283–88.

82. E. G. Campbell et al., "Characteristics of Medical School Faculty Members Serving on Institutional Review Boards: Results of a National Survey," *Academic Medicine* 78 (2003): 831–36; C. H. Coleman, "Rationalizing Risk Assessment in Human Subjects Research," *Arizona Law Review* 46 (2004): 1–51; D. A. Hyman, "Institutional Review Boards: Is This the Least Worst We Can

Do?," *Northwestern University Law Review* 101 (2007): 749–73; Institute of Medicine, *Conflict of Interest in Medical Research, Education, and Practice,* ed. Bernard Lo and Marilyn J. Field (Washington, DC: National Academies Press, 2009); E. Larson et al., "A Survey of IRB Process in 68 U.S. Hospitals," *Journal of Nursing Scholarship* 36 (2004): 260–64; R. McWilliams et al., "Problematic Variation in Local Institutional Review of a Multicenter Genetic Epidemiology Study," *Journal of the American Medical Association* 290 (2003): 360–66; R. G. Spece, D. S. Shimm, and A. E. Buchanan, eds., *Conflicts of Interest in Clinical Practice and Research* (New York: Oxford University Press, 1996).

83. U.S. Food and Drug Administration, Statement of Policy for Regulating Biotechnology Products, 49 Fed. Reg. 50878 (1984); U.S. Food and Drug Administration, Application of Current Statutory Authorities to Human Somatic Cell Therapy Products and Gene Therapy Products, 58 Fed. Reg. 53248 (1993).

84. Letter to the Honorable Joseph Barton, Chairman, Subcomm. on Oversight and Investigation, Comm. on Commerce, U.S. House of Rep., from Diane E. Thompson, Assoc. Comm'r for Legislative Aff., Food & Drug Admin. (Apr. 14, 1995).

85. "Average Cost to Develop A New Biotechnology Product Is $1.2 Billion," *Medical News Today,* Nov. 11, 2006, www.medicalnewstoday.com/articles/56377 .php.

86. Nicholas Thompson, "Gene Blues: Is the Patent Office Prepared to Deal with the Genomic Revolution?," *Washington Monthly,* Apr. 2001, www.washington monthly.com/features/2001/0104.thompson.html.

87. Ass'n for Molecular Pathology v. United States PTO, 2010 U.S. Dist. LEXIS 35418 (S.D.N.Y. Apr. 2, 2010).

88. Gregory Stock, *Redesigning Humans: Our Inevitable Genetic Future* (New York: Houghton Mifflin, 2002), 168.

89. Cornelia Dean, "Scientific Savvy? In U.S., Not Much," *New York Times,* Aug. 30, 2005, www.nytimes.com/2005/08/30/science/30profile.html.

90. Doug Farquhar and Liz Meyer, State Authority to Regulate Biotechnology under the Federal Coordinated Framework, 12 Drake J. of Agric. Law 447 (2007).

91. Lyndsey Layton, "FDA Nears Approval as Food of Genetically Altered Salmon," *Washington Post,* Sept. 7, 2010, A04.

92. Lyndsey Layton, "FDA Rules Won't Require Salmon Labels," *Washington Post,* Sept. 19, 2010, A01.

93. Andrew Zajac, Monica Eng, and Kristin Samuelson, "Panel Tackles Salmon Engineering: One Member Says FDA Will Likely OK Genetically Modified Fish, but Not Soon," *Chicago Tribune,* Sept. 21, 2010, C13.

94. Laura Canter, "Chefs Weigh In on Genetically Modified Salmon," http:// gmwatch.org/latest-listing/1-news-items/12477-opposition-grows-against-gm -salmon.

95. Ibid.

96. Ibid.

97. Franco Furger and Francis Fukuyama, "A Proposal for Modernizing the Regulation of Human Biotechnology," *Hastings Center Report* 37, no. 4 (2007): 16–20.

98. Ibid.

99. Council for Responsible Genetics, "Position Paper on Human Germ Line Manipulation," *Human Therapy* 4 (1992): 35–37.

100. Frankel and Chapman, *Human Inheritable Genetic Modifications*, 23.

101. Ibid., 26.

102. Catherine Wessinger, "New Religious Movements and Violence," in *Introduction to New and Alternative Religions in America*, ed. Eugene V. Gallagher and W. Michael Ashcroft (Westport, CT: Greenwood Press, 2006), 165, 170.

103. Michael Homer, "Children in the New Religious Movements: The Mormon Experience," in Gallagher and Ashcroft, *Introduction to New and Alternative Religions*, 224, 236.

104. Wessinger, "New Religious Movements," 196.

105. DeNeen L. Brown, "The Leader of UFO Land: His Holiness Rael Explains the Origins of Life on Earth. Except for That Clone," *Washington Post*, Jan. 17, 2003, C01.

106. Nick Scott, "Alien Persuasion," *Sunday Telegraph Magazine* (Australia), May 3, 2009, 21.

107. George D. Chryssides, "The Raelian Movement," in Gallagher and Ashcroft, *Introduction to New and Alternative Religions*, 242.

108. George J. Annas, Lori B. Andrews, and Rosario M. Isasi, "Protecting the Endangered Human: Toward an International Treaty Prohibiting Cloning and Inheritable Alterations," *American Journal of Law and Medicine* 28 (2002): 151–78.

109. Prince v. Massachusetts, 321 U.S. 158, 166 (1944).

110. U.S. Department of Health and Human Services, Administration for Children & Families, Child Maltreatment 2008, www.acf.hhs.gov/programs/cb/pubs/cm08/chapter6.htm#removed.

111. Kirsten Rabe Smolensky, "Creating Children with Disabilities: Parental Tort Liability for Preimplantation Genetic Interventions," *Hastings Law Journal* 60 (2008): 299–345.

112. Bonbrest v. Kotz, 65 F. Supp. 138 (D.D.C. 1946).

113. See, e.g., United States of America v. Gary Eugene Straach, 978 F.2d 232 (5th Cir. 1993).

CHAPTER NINE. *Preserving Societal Cohesion*

1. Sharon Beder, *Selling the Work Ethic: From Puritan Pulpit to Corporate PR* (New York: Zed Books, 2000).

2. President's Council on Bioethics, *Beyond Therapy: Biotechnology and the Pursuit of Happiness* (2003), 145.

3. Jack Curry, "4 Indicted in a Steroid Scheme That Involved Top Pro Athletes," *New York Times*, Feb. 13, 2004, A1.

4. G. W. Bush, State of the Union Address (2004), www.whitehouse.gov/news/releases/2004/01/20040120-7.html.

5. Jerome Karabel, "The Legacy of Legacies," *New York Times*, Sept. 13, 2004, A23.

6. Kurt Vonnegut, "Harrison Bergeron," in *Welcome to Monkey House* (New York: Dell, 1998).

7. George J. Annas, "The Man on the Moon, Immortality, and Other Millennial Myths: The Prospects and Perils of Human Genetic Engineering," *Emory Law Journal* 49 (2000): 772–73.

8. S. J. McNamee and R. K. Miller Jr., *The Meritocracy Myth* (Lanham, MD: Rowman and Littlefield, 2004).

9. "The Pulse 2007," *Cleveland Plain Dealer*, Jan. 6, 2007, B9.

10. Charles J. Epstein, "1996 ASHG Presidential Address: Toward the 21st Century," *American Journal of Human Genetics* 60 (1997): 8.

11. Ronald Dworkin, *Sovereign Virtue: The Theory and Practice of Equality* (Cambridge, MA: Harvard University Press, 2002), 445.

12. Jonathan Glover, *What Sort of People Should There Be?* (Hammondsworth, UK: Penguin Books, 1984), 53.

13. Michael Henry, Jennifer Fishman, and Stuart Youngner, "Propranolol and the Prevention of Post-Traumatic Stress Disorder: Is It Wrong to Erase the 'Sting' of Bad Memories?," *American Journal of Bioethics* 7, no. 9 (2007): 12–30.

14. President's Council on Bioethics, *Beyond Therapy*, 154–55.

15. Elaine Wilson, "Children Carry on Family's Military Tradition," *U.S. Air Force*, Feb. 10, 2010, www.af.mil/news/story.asp?id=123189924.

16. Roberto Suro, "Texas Mother Gets 15 Years in Murder Plot," *New York Times*, Sept. 5, 1991, www.nytimes.com/1991/09/05/us/texas-mother-gets-15-years-in-murder-plot.html.

17. National Association of Sports Officials, "Poor Sporting Behavior Incidents," www.naso.org/sportsmanship/badsports.html.

18. National Alliance for Youth Sports, "U.S. Youth Sports Parents Seen as World's Worst, Poll Finds," Apr. 4, 2010, www.nays.org/fullstory.cfm?articleid=10389.

CHAPTER TEN. *Providing for Our Descendents*

1. David Sloan Wilson, *Evolution for Everyone: How Darwin's Theory Can Change the Way We Think about Our Lives* (New York: Delacort Press, 2007), 314.

2. U.S. Food and Drug Administration, Guidance for Industry and Other Stakeholders, "Toxicological Principles for the Safety Assessment of Food Ingredients, Guidelines for Reproductive Studies" (Redbook, 2000), www.fda.gov/Food/GuidanceComplianceRegulatoryInformation/GuidanceDocuments/FoodIngredientsandPackaging/Redbook/ucm078396.htm.

3. U.S. Food and Drug Administration, Database of Select Committee on GRAS

Substances, "Caffeine," www.accessdata.fda.gov/scripts/fcn/fcnDetailNavigation .cfm?rpt=scogsListing&id=42.

4. A. Velimirov, C. Binter, and J. Zentek, "Biological Effects of Transgenic Maize NK603xMON810 Fed in Long Term Reproduction Studies in Mice," Report, Forschungsberichte der Sektion IV, Band 3, Institut für Ernährung, and Forschungsinttitut für biologischen Landbau, Vienna, Austria, Nov. 2008, http://fbae.org/2009/FBAE/website/false-propaganda_gm_maize_reduces_fertility_deregulates_genes_in_mice_dr_mae_wan_ho_isis.html.

5. U.S. Food and Drug Administration, Research Highlights, "Multigenerational Studies of Low-Dose Ethinyl Estradiol," Apr. 2007, www.fda.gov/About FDA/CentersOffices/NCTR/WhatWeDo/ucm071383.htm.

6. Ibid., 21.

7. Fosamax TV Ad (2004), www.youtube.com/watch?v=dqhkfQHw_G4.

8. Boniva TV Ad (2007), http://wn.com/Boniva_TV_Ad_2007.

9. Gina Kolata, "When Drugs Cause Problems They Are Supposed to Prevent," *New York Times*, Oct. 17, 2010, 17.

10. Ibid., 23.

11. Pate v. Threlkel, 661 So.2d 278 (Fla. 1995).

12. Safer v. Pack, 677 A.2d 1188 (N.J. Super. Ct. App. Div. 1996).

13. Susannah Baruch et al., "Human Germline Genetic Modifications: Issues and Options for Policymakers" (Genetics and Public Policy Center, 2005), 44.

14. World Transhumanist Association, "Position on Human Germline Genetic Modification" (2005), www.transhumanism.org/index.php/WTA/more/636/.

15. Gail Javitt, "Drugs and Vaccines for the Common Defense: Refining FDA Regulation to Promote the Availability of Products to Counter Biological Attacks," *Journal of Contemporary Health Law and Policy* 19 (2002): 37–116.

16. Ibid.

17. N. Cahn, "Necessary Subjects: The Need for a Mandatory National Donor Gamete Registry," *DePaul J. Health Care Law*, 2008, 203–24.

18. Marc Lappé, "Ethical Issues in Manipulating the Human Germline," *Journal of Medicine and Philosophy* 16 (1991): 621–39.

CHAPTER ELEVEN. *Safeguarding the Human Species*

1. 21 CFR § 46.111(a)(2).

2. Ian Sample, "Craig Venter Creates Synthetic Life Form," *Guardian* (UK), May 20, 2010, www.guardian.co.uk/science/2010/may/20/craig-venter-synthetic -life-form.

3. U.S. Congress, Office of Technology Assessment, "Human Gene Therapy—a Background Paper," OTA-BP-BA-32 (Washington, DC: U.S. Government Printing Office, 1984).

4. "Japan Bans Seed from Setting Up Cloning Research Center," *Transplant News*, Jan. 15, 1999, www.allbusiness.com/health-care-social-assistance/ambulatory-health-services/151398-1.html.

5. Jane Barrett, "Doctor Vows Cloning Effort Will Proceed," *Philadelphia Inquirer*, Aug. 7, 2001, A02.

6. Melissa Healy, "Sold on Drugs, Wooing the Gatekeeper," *Los Angeles Times*, Aug. 6, 2007, F3.

7. Jonathan Glover, *What Sort of People Should There Be?* (Hammondsworth, UK: Penguin Books, 1984), 51.

8. Mark S. Frankel and Audrey R. Chapman, *Human Inheritable Genetic Modifications: Assessing Scientific, Ethical, Religious, and Policy Issues* (Washington, DC: AAAS Publications Services, 2000), 34, www.aaas.org/spp/sfrl/projects/germline/report.pdf

9. Peter Tyson, "A Short History of Quarantine," *NOVA*, PBS, 2004, www.pbs.org/wgbh/nova/body/short-history-of-quarantine.html.

10. Stephanie Strom, "Local Laws Fighting Fat under Siege," *New York Times*, June 30, 2011, www.nytimes.com/2011/07/01/business/01obese.html?_r=1&ref=transfattyacids.

11. "LADWP 'Opens the Pipes' on New System to Replace City's Largest, Oldest Water Line," *Business Wire*, Aug. 4, 2003, www.allbusiness.com/energy-utilities/utilities-industry-water/5745964-1.html.

12. Centers for Disease Control and Prevention, Mission Statement, www.cdc.gov/maso/pdf/cdcmiss.pdf.

13. Lawrence O. Gostin, Public Health Law: Power, Duty, Restraint 25–41 (2000); Gonzales v. Raich, 545 U.S. 1 (2005).

14. Paul A. Lombardo, Three Generations, No Imbeciles: New Light on *Buck v. Bell*, 60 N.Y.U.L. Rev. 30, 49–62 (1985).

15. Buck v. Bell, 274 U.S. 200 (1927).

16. U.S. Department of Health and Human Services, "Fact Sheet on the 1946–1948 U.S. Public Health Service Sexually Transmitted Diseases (STD) Inoculation Study," www.hhs.gov/1946inoculationstudy/factsheet.html.

17. Tyson, "Short History of Quarantine."

18. Oregon State University, "It's in the Blood: A Documentary History of Linus Pauling, Hemoglobin, and Sickle-Cell Anemia," http://osulibrary.oregonstate.edu/specialcollections/coll/pauling/blood/narrative/page35.html.

19. Wendy E. Parmet, "Legal Power and Legal Rights—Isolation and Quarantine in the Case of Drug-Resistant Tuberculosis," *New England Journal of Medicine* 357, no. 5 (2007): 433–35.

20. Meredith Hobbs, "Fulton County Daily Report," Sept. 13, 2007, www.law.com/jsp/law/sfb/lawArticleSFB.jsp?id=1189587766087.

21. Richard Knox, "Arizona TB Patient Jailed as a Public Health Menace," National Public Radio, June 11, 2007, www.npr.org/templates/story/story.php?storyId=10874970.

22. New York Times Topics, "Joseph M. Arpaio," Dec. 15, 2011, topics.nytimes.com/topics/reference/timestopics/people/a/joseph_m_arpaio/index.html.

23. "ACLU of Arizona Sues County Officials over Inhumane Confinement of

TB Patient," American Civil Liberties Union, May 31, 2007, www.aclu.org/tech
nology-and-liberty/aclu-arizona-sues-county-officials-over-inhumane-confine
ment-tb-patient.

24. Ibid.

25. Turning Point Model State Public Health Act of 2003, §§ 5-106(d)(1),
1–102(6), www.turningpointprogram.org/Pages/pdfs/statute_mod/MSPHAfinal
.pdf.

26. Ibid., § 5-106(d)(2).

27. "Glenn Beck Claims Science Czar John Holdren Proposed Forced Abortions and Putting Sterilants in the Drinking Water to Control Population,"
St. Petersburg Times Politifact, July 29, 2009, http://politifact.com/truth-o-meter/
statements/2009/jul/29/glenn-beck/glenn-beck-claims-science-czar-john
-holdren-propos/.

28. Paul R. Ehrlich, Anne H. Ehrlich, and John P. Holdren, *Ecoscience: Population, Resources, Environment* (San Francisco: W. H. Freeman, 1977), 788.

29. Ibid., 783.

30. Ibid., 786.

31. Ibid., 788.

32. Gonzales v. Carhart, 550 U.S. 124 (2007).

33. Ehrlich, Ehrlich, and Holdren, *Ecoscience*, 837.

34. Ibid., 838.

35. Monica Sharma, "Twenty-First Century Pink or Blue: How Sex Selection
Technology Facilitates Gendercide and What We Can Do about It," *Family Court
Review* 46 (2008): 198.

36. Drew Christiansen, "Ethics and Compulsory Population Control," *Hastings Center Report* 7, no. 1 (1977): 30–33.

37. Lexi Krock, "Population Campaigns," *NOVA*, PBS, 2004, www.pbs.org/
wgbh/nova/worldbalance/campaigns.html.

38. Ehrlich, Ehrlich, and Holdren, *Ecoscience*, 767.

39. Ibid., 768.

40. Rohinton Mistry, *A Fine Balance* (New York: Vintage Books, 1995),
521–25.

41. Laura Fitzpatrick, "China's One-Child Policy," *Time*, July 27, 2009, www
.time.com/time/printout/0,8816,1912861,00.html.

42. Ehrlich, Ehrlich, and Holdren, *Ecoscience*, 772.

43. Fitzpatrick, "China's One-Child Policy."

44. "China: One-Child Policy Will Stand," CNN, Sept. 27, 2010, http://arti
cles.cnn.com/2010-09-27/world/china.one.child.policy_1_family-planning
-policy-family-planning-commission-child?_s=PM:WORLD.

45. Frances Fukuyama, *The End of History and the Last Man* (New York:
Avon Books, 1992).

INDEX

against persons who are not genetically engineered, 28–29
Down syndrome, 165, 174, 221
Dworkin, Ronald, 97, 162, 198

economic competition among nations. *See* international economic competition, role in genetic engineering
education, role of genes in, 17
education, access to, and genetic engineering, 117–18, 194
egg donation. *See* gamete (egg and sperm) donation
Ehrlich, Paul and Anne, 223–24, 226, 228. *See also* population growth
Emanuel, Ezekiel, 178
embryos (human), legal protections for, 156, 169, 172, 174–75, 188, 225; in research, 59, 82, 176–77, 179. *See also* Dickey-Wicker Amendment; stem cells
employment, role in decisions about genetic engineering, 117
environment: effect of, on genes, 136; effect of, on human lineage, 127, 129, 131, 223, 225; Jean-Baptiste Lamarck, 136; role in evolution, 133, 135; role in human health and behavior, 69
Environmental Protection Agency (EPA), 40, 44, 55, 182; and StarLink corn, 56–57, 182
epigenetics, 69, 112, 136
equality: effect of genetic engineering on, 85, 92–94, 101–2, 170, 192, 195; lack of, in U.S. society, 195
eugenics, 100, 121–24, 167, 169, 223; and *Buck v. Bell*, 121, 216; Sweden, 122. *See also* Nazis; sterilization
evo-devo, 69, 149–51

Faden, Ruth, 176
Fagan, John B., 128
Falk, Dean, 142, 144, 145, 146, 150
Falwell, Jerry, 217
fertility industry, 3, 77, 130, 175–76; liability of clinics to genetically engineered children, 190; profitability of, 182; regulation of, to protect lineage, 214; in the United Kingdom, 174
fetuses, human: legal protections for, 225; protections for use in research, 177, 178
Flavr Savr tomato, 54–55
Fletcher, Joseph, 5, 87, 96–97, 137
Food and Drug Administration, 14, 20, 35, 40, 117, 124, 202; and AIDS, 205–6; and animal testing to establish safety of genetic engineering in humans, 201–2; and AquAdvantage salmon, 183–84; and death of Jessie Gelsinger, 64; and off-label uses of approved products, 125, 210; and practice of medicine, 179, 183; regulation of genetically modified food, 182; regulation of genetic engineering, 124–25, 157, 183, 185; regulation of human subjects research, 179–80. *See also* biotechnology industry marketing; gene therapy
fossils, 134–36, 140–41, 144–46, 147. *See also* paleontology
Fukuyama, Frances, 85–86, 184, 228

gamete (egg and sperm) donation, 3, 12, 58, 60, 74, 116, 174; lawsuits involving, 77–78, 99; payment for, 175–76; protections for women, 175–76; tracking children born following, 206
Garreau, Joel, 13, 15–16, 17, 19
Gattaca, 28–29
Gelsinger, Jessie, 64, 161
gender. *See* sex selection
gene flow, 133, 134
gene therapy, 4, 12, 21, 22, 59, 124, 158; American Medical Association guidelines on, 116; and death of Jessie Gelsinger, 63–64; and French immune system experiment, 64–65; and Unabomber, 100
genetically modified food: AquAdvantage salmon, 183–84; extent of use, 57; opposition to, 55, 71, 115, 206; regulation of, 182–84; risks from, 55–57. *See also* StarLink corn

international treaties, 76, 155, 191. *See also* germ line genetic engineering: international regulation of
in vitro fertilization, 3, 12, 28, 58, 77–78, 87, 88, 118, 125, 162, 174, 181; and Louise Brown, 58, 88, 89, 211; constitutional protections for, 167; cost of, 75, 196; regulation of, 156, 225; risks of, 66; and "wrongful life" lawsuits, 162–63. *See also* genetic testing: preimplantation; reproductive freedom

Javitt, Gail, 205
Johanson, Donald, 141
Judaism, 118–19

Kass, Leon, 6, 80, 84, 97, 132, 130, 184; and commodification of children, 77; and in vitro fertilization, 87; objections to human cloning, 75; objections to performance enhancement in sports, 192; objection to propranolol, 198; and "yuck factor," 81
Kaczynski, Ted (Unabomber), 100
Kolata, Gina, 202
Kress, Nancy, *Beggars in Spain*, 26–28, 32, 73, 81, 91
Kurzweil, Ray, 18, 23, 111

lactose intolerance, 95
Lamarck, Jean-Baptiste, 136
Lappé, Marc, 206–7
lawsuits: prenatal harm, 189; "wrongful life," 163–64
Leakey, Richard, and Martin Pickford, 140–41, 145
legality of genetic engineering, 119
Lombardo, Paul, 216–17
longevity. *See* anti-aging

marijuana, regulation of, 214
mating rituals, human, 174; and computer dating, 2, 156–57; and creation of separate human species, 105; and genetic testing, 2–3
Mayr, Ernst, 103, 109–10, 130–31, 135; and bipedalism, 145; and evolutionary progress, 137–38; and size of human brain, 150

McKibben, Bill, 75
Medicaid, 215, 221
medical malpractice, 163–64, 203–4; prenatal harm lawsuits, 189; "wrongful life" lawsuits, 163–64
Mendel, Gregor, 133
military, genetic engineering in the, 15, 120–21, 156, 199, 212; and diminishing moral judgment, 198–99
"missing link," 137, 141
Mistry, Rohan, *A Fine Balance*, 226–27. *See also* sterilization
mitochondrial DNA, 148–49
modern synthesis. *See* Huxley, Julian
molecular biologists vs. paleontologists, 147–49
molecular genetics, 135
morality: diminishing, 197–200; enhancing, 19; propranolol and post-traumatic stress disorder (PTSD), 198
Mormons, 186
multiregional hypothesis. See *Homo erectus*
myostatin, 15

Naam, Ramez, 13, 18, 19
nanotechnology, 35, 43–46, 196; and Richard Drexler, 44–46; and gray goo, 45–46
National Academy of Sciences, 82–83, 179; policy on egg donation, 176
National Institutes of Health, 20, 35, 64, 65, 66, 82, 97, 116, 117, 126; Clinical Center, 178; and consideration of long-term effects of human subjects research, 210; Recombinant DNA Advisory Committee (RAC), 35, 180; sanctioning Martin Cline, 211. *See also* Collins, Frances; Human Genome Project
natural selection, 131, 133, 134, 138
Nazis, 121, 122, 212, 229
Neanderthals, 102, 107–10, 146, 149
newborn screening, 123–24, 223; to detect diminished moral judgment, 199; to detect illegal genetic engineering, 214, 222, 223; religious objections to, 124

Nozick, Robert, 126
nuclear war, 36; and destruction of
humanity, 36

Olympics, 14, 43, 95, 170–71. *See also*
sports
ovarian hyperstimulation syndrome, 176

paleontology, 138, 139, 145; vs. molecular biologists, 147–49
parental decision making, 73–76; and
ability to assess risks of genetic engineering, 181; and "best interests" test,
167–68; and children's rights, 78–79,
165, 166, 172, 173; liability to children
for adverse effects of, 188; and loss
of genetic diversity, 116, 222; and
reasonable person standard, 172–73;
and sports, 166, 170–71; about their
children's health, 165–66, 167, 171;
and use of children as research subjects, 171, 175, 203. *See also* abuse
and neglect laws; cults; morality;
reproductive behavior; reproductive
freedom; sex selection
Parfit, Derek, Non-Identity Problem,
161–65
particle accelerators and destruction of
humanity, 36–39, 210
patents, 40, 180–81, 184, 201
Pauling, Linus. *See* sickle cell disease
performance enhancement in sports,
14–15, 20
Pickford, Martin. *See* Leakey, Richard
Piltdown man, 137
Planned Parenthood, 123
pleiotropy, 67, 68, 127
population genetics, 135–36
population growth, 223–24; Chinese
response to, 227–28; Indian response
to, 226–27. *See also* Ehrlich, Paul and
Anne; Holdren, John; sterilization
Posner, Richard, 34–35, 41, 46, 47
pregnancy in transgender men, 99
pre-implantation genetic testing and
diagnosis, 3, 28, 58–59, 75, 80, 174.
See also genetic testing; in vitro fertilization
prenatal harm lawsuits, 189
products liability, 190

public health law: Centers for Disease
Control and Prevention, 214; constitutional basis for, 213, 214–17,
224–25; and contact tracing, 213;
federal vs. state exercise of, 214; and
model act, 221; and obesity, 213; and
polio, 215; and protecting human
lineage, 213–14; and quarantine, 213,
220–21; and response to 9/11, 221;
and sanitation systems, 213, 214, 215;
scope of powers, 213, 221; and smallpox, 215–16; and syphilis testing, 217;
and tax credits for compliant parents,
222; and tuberculosis, 213, 219–21;
and U.S. Public Health Service experiment at Tuskegee, 177–78, 217; and
U.S. Public Health Service experiment
in Guatemala, 217. *See also* eugenics;
HIV testing; sickle cell disease

random mutation, role in evolution,
133, 134
recombinant DNA, 14, 20, 39–40, 59,
201; in animals, 54–57; fears about,
40–41; Flavr Savr tomato, 54–55;
Recombinant DNA Advisory Committee (RAC), 180; research moratorium, 41
regulatory DNA, 67
religion: Catholic Church, views on
genetic engineering, 6; and evolution,
132, 137; and genetic engineering,
6, 24, 26, 75, 96–97, 118; Judaism,
118–19; and newborn screening, 124;
and public health measures
to protect lineage, 225. *See also*
creationism
replacement hypothesis. See *Homo
erectus*
reproductive behavior: to create genetically engineered children, 87; and
inability to mate with non-engineered
individuals, 104–5, 212, 222; by persons who were genetically engineered,
25, 27, 88, 104; and refusal to mate
with non-engineered individuals, 101,
105–6, 209; and reproductive success,
128–29, 135; in science fiction, 30–32.
See also reproductive freedom; sex
selection; sterilization

reproductive freedom, 166–67, 211–12, 224–25; and assisted reproductive technologies, 67; in China, 227–28; in India, 226–27; and permit requirement, 223. *See also* population growth; sterilization
reproductive isolation, behavioral. *See* reproductive behavior
reproductive isolation, mechanical. *See* reproductive behavior
research, human subjects: Belmont Report, 178; and Martin Cline, 211; Common Rule and regulation of, 178–79, 209–10; government regulations on children in, 171; IRB consideration of long-range effects of research, 209–10; NIH refusal to fund genetic enhancement, 97, 175; and private sector, 175, 179–80; on race and intelligence, 210; role of researchers in preventing anti-science backlash, 197; state government funding of, 175. *See also* embryos (humans), legal protection for; parental decision making
Robertson, John, 162
Roe v. Wade, 166

Sandel, Michael, 75
savannah hypothesis, 141–45
Savulescu, Julian, 6, 16, 23, 79, 81, 118, 131
sex selection, 18, 74, 174, 222; impact on human lineage, 74, 129–30, 208, 210, 223; in India, 74; regulation outside of the U.S., 174, 225; in the United Kingdom, 174
sexual abuse, 168; by cults, 186, 187–88
sickle cell disease, 68; and Linus Pauling, 219; testing programs for, 218–19
Silver, Lee, 12–13, 16, 17
Smolensky, Kirsten Rabe, 188
Speaker, Andrew, 219–20
sperm donation. *See* gamete (egg and sperm) donation
sports: and diminishing moral judgment, 199; genetic engineering in, 119–20; overcoming competitive advantages, 193; performance enhancement in, 125, 192, 193, 194;

risks to children from participating in, 170; and steroids, 192. *See also* genetic enhancement; human growth hormone; Olympics; parental decision making
StarLink corn, 56–57, 182
stem cells: research using, 22, 23, 82–83, 97, 159, 174–75; state programs, 214; use in genetic engineering, 22
sterilization, 121, 122, 167, 216, 223; in China, 228; in India, 226–27; Rohan Mistry, 227–27; to prevent multigenerational harms, 206–7; as punishment for parents who diminish moral judgment, 200; use of, to control population growth, 223, 226–27. *See also* eugenics
Stock, Gregory, 13, 17, 18, 74; on germ line genetic engineering, 12, 61; on parental decision making, 181; on pre-implantation genetic testing, 59; on speciation, 103; transhumanist vision, 5, 6
Sturgeon, Theodore, 104
Sunstein, Cass, 34, 35, 46; and precautionary principle, 46, 151, 209
Sweden, 122
syphilis. *See* public health law

Tattersall, Ian, 138
Tay-Sachs disease. *See* genetic disease
transhumanism, 6, 23–24, 96, 106, 132, 158, 204; cults, 119; democratic, 24; and Humanity+, 24; libertarian, 24; and Mormons, 24; and relaxing regulation of genetic engineering, 204–5, 206; resemblance to religion, 24; and World Transhumanist Association, 24, 204; and World Transhumanist Society, 24. *See also specific transhumanists*
tuberculosis. *See* public health law

Unborn Victims of Violence Act, 168
United Kingdom, regulation of: animal-human hybrids, 156, 159, 177; sex selection, 174; stem cell research, 159
U.S. Public Health Service. *See* public health law

valuing human lives, methods for, 47–48
Venter, Craig, 211
Vonnegut, Kurt, *Harrison Bergeron*, 194–95

Wade, Nicholas, 129, 138
Ward, Peter, 5, 53, 61, 63, 103, 105–6
Watson, James, 18, 131

Wells, H. G., 102, 195
Wilson, David Sloan, 201
Wilson, Edmund O., 198
"wrongful life" lawsuits, 163–64

xenotransplantation, 83

Yao Ming, 119–20. *See also* China
Young, Simon, 11, 13, 17, 21, 24